MEMOIRS
OF THE
AMERICAN MATHEMATICAL SOCIETY

Number 117

EXTENSIONS OF POSITIVE-DEFINITE FUNCTIONS

by

JOHN R. McMULLEN

Published by the
American Mathematical Society
Providence, Rhode Island
1972

International Standard Book Number 0-8218-1817-1
Library of Congress Catalog Number 52-42839
AMS 1970 Subject Classifications: 22D10, 22D99, 43A35, 43A60, 43A65, 43A80; 43A07, 43A05, 43A15, 43A22

Printed in the United States of America

EXTENSIONS

OF

POSITIVE-DEFINITE FUNCTIONS

John R. McMullen[1]

Institute for Advanced Study
Princeton, N. J. 08540

Received November 3, 1971

---

[1]Supported in part by National Science Foundation Grant GP7952X3. Most of the work herein presented is contained in the author's doctoral dissertation written at the University of Washington under the guidance of Professor E. Hewitt.

Received by the editors November 3, 1971.

Abstract. Conditions are sought under which a continuous positive-definite function on a closed subgroup is a restriction of such a function defined on the whole group. This is found to be true for an arbitrary compact subgroup of a locally compact group, for an arbitrary open subgroup, for an almost-periodic continuous positive-definite function provided the Bohr topology on the subgroup is that induced by the Bohr topology of the whole group, for an arbitrary subgroup of any group with a closed locally compact abelian subgroup of finite index, and for positive-definite functions in the Fourier Algebra. It is also shown to be true for the center of a group with small invariant neighborhoods, and evidence is collected to support a conjecture that the same is true for the center of an arbitrary locally compact group. Truth of this conjecture is equivalent to truth of the converse of an often used result of Douady.

Key words and phrases. Locally compact group, extension problem, positive-definite function, unitary representation, almost periodic function, Bohr compactification, Bohr topology, Fourier algebra, center, central extension property.

CONTENTS

Section

§1. INTRODUCTION

We are concerned in this paper with answering the following question. Let G be a topological group, and let G_o be a closed subgroup of G. Let φ_o be a positive-definite continuous function defined on G_o. What conditions on G, G_o and φ_o will ensure the existence of a positive-definite continuous function φ on G that extends φ_o?

(1.1) Definition. Let G, G_o and φ_o be as above. If φ_o has a positive-definite continuous extension φ on G, we shall say that (G,G_o,φ_o) is an *extension triple*. If (G,G_o,φ_o) is an extension triple for all closed subgroups G_o of G and all positive-definite continuous functions φ_o on G_o, we shall say that the group G has the *extension property*. Given a topological group G, a closed subgroup G_o of G such that (G,G_o,φ_o) is an extension triple for all positive-definite continuous functions φ_o on G_o will be called an *extending subgroup* of G.

(1.2) Discussion. The literature contains several results of interest in this connection. It is known, for example, that all locally compact abelian groups and all compact groups have the extension property (Hewitt and Ross [2,(34.48)]), and that every open subgroup of a locally compact group G is extending (Hewitt [1], for example). Thus we see that every discrete group has the extension property.

Numerous partial results are also known. Kaniuth [1] proves that if G is a locally compact group whose conjugacy classes have compact closure and which has the small invariant neighborhood property, if G_o is a compact normal subgroup of G, and if $\varphi_o(y^{-1}xy) = \varphi_o(x)$ for all $y \in G$ and all $x \in G_o$, then (G,G_o,φ_o) is an extension triple. Grosser and Moskowitz [1] prove that every compact subgroup and every closed subgroup of the centre $Z(G)$ of a central topological group G is an extending subgroup.

Our contributions to this list are the following. We have completely generalized the above result of Kaniuth, and shown that every compact subgroup of a locally compact group is extending (see (2.7) and (4.22) infra). Every locally compact group G containing a closed subgroup H of finite index with the extension property has the extension property (see (3.7) infra); in particular, every locally compact group G with a closed abelian subgroup of finite index has the extension property.

We also derive several partial results. In §4 we show that (G,G_o,φ_o) is an extension triple whenever φ_o belongs to the Fourier algebra of G_o. In §5 we show that $(G,Z(G),\varphi_o)$ is an extension triple whenever G has the small invariant neighborhood property, a significant generalization of a result of Grosser and Moskowitz quoted above.

Finally, we have proved a number of related results which illuminate the main question, though perhaps less directly. Into this category fall the results of §2 concerning almost periodic positive-definite continuous functions, certain questions of "permanence" raised in §3, and several results from §5. We have also included examples which we hope will help guide future conjectures concerning this extension question. A conjecture of our own appears in §5.

(1.3) Notation and Terminology. Throughout this treatise all unexplained symbols, terms, and conventions are defined explicitly in Hewitt and Ross [1] and [2]. In addition we agree that "representation" shall mean "continuous unitary representation", since it is only with the latter concept that we deal.

Some of our usage is not to be found in loc. cit.. The function f is called Hermitian if $f^\star = \overline{f}$. The representation space of a representation U we shall denote by $H(U)$. Neighborhoods are not necessarily open.

(1.4) Definition. Let G be a topological group, and let G_o be a closed subgroup of G. Let U and $U^{(o)}$ be representations of G and G_o respectively. We shall say that U extends $U^{(o)}$ if $H(U^{(o)})$ is contained in $H(U)$ and if

(1) $$U_x\xi = U_x^{(o)}\xi$$

for all $x \in G_o$ and all $\xi \in H(U^{(o)})$. In the special case where $G = G_o$ we shall say that $U^{(o)}$ is a subrepresentation of U.

(1.5) Theorem. Let G_o be a closed subgroup of the topological group G, and let φ_o be a positive-definite continuous function defined on G_o. Then φ_o has a positive-definite continuous extension φ on G if and only if the corresponding representation $V^{(\varphi_o)}$, defined as in Hewitt and Ross [2,(32.3)], has an extension to G in the sense of (1.4) supra. This extension may be taken to be equivalent to $V^{(\varphi)}$.

Proof. Suppose that $\zeta_o \in H(V^{(\varphi_o)})$ is some cyclic vector for $V^{(\varphi_o)}$ such that for $x \in G_o$ we have

$$\varphi_o(x) = < V_x^{(\varphi_o)} \zeta_o, \zeta_o > ,$$

and suppose that there is a representation V extending $V^{(\varphi_o)}$ in the sense of (1.4). Then we may define $\varphi(x)$ for $x \in G$ by setting

$$\varphi(x) = < V_x\zeta_o, \zeta_o > .$$

The function φ evidently is a positive-definite continuous extension of φ_o.

Conversely, suppose that φ_o has an extension φ to G that is positive-definite and continuous. Let ζ and ζ_o be cyclic vectors for $V^{(\varphi)}$ and $V^{(\varphi_o)}$ such that

$$\varphi(x) = < V_x^{(\varphi)}\zeta, \zeta >$$

for $x \in G$, and

$$\varphi_o(x) = \langle V_x^{(\varphi_o)} \zeta_o, \zeta_o \rangle$$

for $x \in G_o$. Since φ extends φ_o, we have

$$\langle V_x^{(\varphi_o)} \zeta_o, \zeta_o \rangle = \varphi_o(x) = \langle V_x^{(\varphi)} \zeta, \zeta \rangle$$

for $x \in G_o$. Hewitt and Ross [2,(32.8.b)] tell us that therefore $V^{(\varphi_o)}$ is equivalent to the representation V' obtained by restricting all operators $V_x^{(\varphi)}$ to the closure H' in $H(V^{(\varphi)})$ of the linear span of the set $\{V_x^{(\varphi)}\zeta : x \in G_o\}$. Thus $V^{(\varphi_o)}$ is equivalent to a representation V' of which $V^{(\varphi)}$ is an extension to all of G. This proves the claim.

§2. EXTENSION OF AN ALMOST PERIODIC POSITIVE-DEFINITE CONTINUOUS FUNCTION

In this section we take up the following question. Given a locally compact group G and a closed subgroup G_o of G, what conditions on G and G_o will ensure that every almost periodic positive-definite continuous function on G_o will have an almost periodic positive-definite continuous extension to all of G?

(2.1) Theorem. A nonzero positive-definite continuous function φ on a topological group G is almost periodic if and only if the corresponding representation $V^{(\varphi)}$ is a direct sum (possibly infinite) of finite dimensional representations.

Proof. Let $G \xrightarrow{\beta} bG$ be the canonical homomorphism from G into its Bohr compactification bG.[1]

[1] For a definition and relevant properties of the Bohr compactification, see Alfsen and Holm [1].

Suppose first that φ is almost periodic, and let ${}^b\varphi$ be the unique continuous function on bG such that $\varphi = ({}^b\varphi) \circ \beta$. Let $\alpha_1, \ldots, \alpha_m$ be complex numbers and let $x_1, \ldots, x_m \in G$. Then we have

$$\begin{aligned} \sum_{j=1}^{m} \sum_{k=1}^{m} \bar{\alpha}_j \alpha_k {}^b\varphi(\beta(x_j)^{-1}\beta(x_k)) &= \sum_{j=1}^{m} \sum_{k=1}^{m} \bar{\alpha}_j \alpha_k {}^b\varphi(\beta(x_j^{-1}x_k)) \\ &= \sum_{j=1}^{m} \sum_{k=1}^{m} \bar{\alpha}_j \alpha_k \varphi(x_j^{-1}x_k) \\ &\geq 0, \end{aligned}$$

and we see that the restriction of ${}^b\varphi$ to the subgroup $\beta(G)$ of bG is positive-definite.

Let $m \geq 1$ and let $\alpha_1, \ldots, \alpha_m$ be complex numbers. The mapping

$$(*) \qquad (y_1, y_2, \ldots, y_m) \longmapsto \sum_{j=1}^{m} \sum_{k=1}^{m} \bar{\alpha}_j \alpha_k ({}^b\varphi)(y_j^{-1}y_k)$$

is clearly a continuous mapping from G^m into K. The set

$$P = \left\{ (y_1, y_2, \ldots, y_m) \in (bG)^m : \sum_{j=1}^{m} \sum_{k=1}^{m} \bar{\alpha}_j \alpha_k {}^b\varphi(y_j^{-1}y_k) \geq 0 \right\}$$

is therefore closed in $(bG)^m$. The previous paragraph shows that P contains $\beta(G)^m$, and hence the closure of $\beta(G)^m$ in $(bG)^m$. Since $\beta(G)$ is dense in bG, we thus have $P = (bG)^m$, so that

$$\sum_{j=1}^{m} \sum_{k=1} \bar{\alpha}_j \alpha_k ({}^b\varphi)(y_j^{-1}y_k) \geq 0$$

for all sequences $(\alpha_1, \ldots, \alpha_m)$ of complex numbers and all sequences $(y_1, \ldots, y_m)$ of elements of bG. Thus ${}^b\varphi$ is positive definite.

Consider the representation $V^{({}^b\varphi)}$ of bG. Since bG is compact, $V^{({}^b\varphi)}$ is a direct sum of finite-dimensional representations. Define a continuous unitary representation U of G by setting $U_x = V^{(^b\varphi)}_{\beta(x)}$ for all $x \in G$. Then U is also a direct sum of finite-dimensional representations, and for all $x \in G$ we have

$$< U_x\zeta, \zeta > = < V^{(^b\varphi)}_{\beta(x)}\zeta, \zeta > = (^b\varphi)(\beta(x))$$
$$= \varphi(x),$$

where ζ denotes the canonical cyclic vector for $V^{(^b\varphi)}$. Hewitt and Ross [2,(32.8.b)] allows us to conclude that the representation U, with all operators U_x restricted to the closed subspace of $H(V^{(^b\varphi)})$ spanned by $\{U_x\zeta : x \in G\}$ is equivalent to $V^{(\varphi)}$. This closed subspace is in fact all of $H(V^{(^b\varphi)})$; for the mapping $y \longmapsto V^{(^b\varphi)}_y$ from bG into $H(V^{(^b\varphi)})$ is continuous, and therefore the set $\{U_x\zeta : x \in G\} = \{V^{(^b\varphi)}_{\beta(x)}\zeta : x \in G\}$ is dense in the set $\{V^{(^b\varphi)}_y\zeta : y \in bG\}$, whose closed linear span is $H(V^{(^b\varphi)})$. Thus U and $V^{(\varphi)}$ are equivalent, and hence $V^{(\varphi)}$ has the required direct sum decomposition.

Conversely, suppose $V^{(\varphi)}$ has a decomposition $\bigoplus_{\iota\in I} U_\iota$ with all U_ι finite-dimensional. Then for each $\iota\in I$, there is a uniquely determined finite-dimensional representation $^bU_\iota$ of bG such that we have $(U_\iota)_x = (^bU_\iota)_{\beta(x)}$ for all $x \in G$. Set $^bU = \bigoplus_{\iota\in I} {^bU_\iota}$. Then we have $^bU_{\beta(x)} = V^{(\varphi)}_x$ for all $x \in G$, and we see as before that

$$\varphi(x) = < V^{(\varphi)}_x\zeta, \zeta >$$

$$= < {^bU_{\beta(x)}}\zeta, \zeta >$$

$$= \psi(\beta(x))$$

where we have set $\psi(y) = < {^bU_y}\zeta, \zeta >$ for $y \in bG$. Thus $\varphi = \psi \circ \beta$, and ψ is continuous and positive definite on bG, so φ is almost periodic, continuous, and positive definite on G.

(2.2) Discussion and notation. Let G_1 and G_2 be topological groups, and let f be a continuous homomorphism from G_1 into G_2. Let bG_j denote the Bohr compactification of G_j and let β_j be the canonical homomorphism of G_j into bG_j $(j = 1,2)$.

The continuous homomorphism $\beta_2 \circ f$ maps G_1 into bG_2, and the closure of $(\beta_2 \circ f)(G_1)$ in bG_2 is compact. Therefore, by the universal property enjoyed by the mapping β_1, there exists a continuous homomorphism, which we shall denote by $b(f)$, from bG_1 into bG_2, such that $b(f) \circ \beta_1 = \beta_2 \circ f$.

(2.3) Theorem. Let G be a topological group, and let G_o be a closed subgroup of G. The following conditions are equivalent.

(i) Every almost periodic positive-definite continuous function on G_o has an almost periodic positive-definite continuous extension on G.

(ii) Every finite-dimensional irreducible representation of G_o extends to a finite-dimensional irreducible representation of G.

(iii) The map $bG_o \xrightarrow{b(j)} bG$, where j is the natural injection of G_o into G, is 1-1, and hence an isomorphism of bG_o into bG.

Proof. We adopt the schema (i) $\Longrightarrow$ (ii) $\Longrightarrow$ (iii) $\Longrightarrow$ (i).

Suppose that (i) holds.[1] Let U be a finite-dimensional irreducible representation of G_o. Its characteristic function $\varphi_o(x) = \langle U_x\zeta, \zeta \rangle$ is an almost periodic positive-definite continuous function on G_o. Since φ_o extends to G as an almost periodic positive-definite continuous function φ, theorems (1.5) and (2.1) show that $V^{(\varphi)}$ is an extension of U and is a direct sum of finite-dimensional representations. Let $\bigoplus_{\alpha\in A} H_\alpha$ be a decomposition of $H(V^{(\varphi)})$ into finite-dimensional subspaces H_α upon each of which $V^{(\varphi)}$

[1]The proof of the implication (i) $\Longrightarrow$ (ii) follows closely the proof of Hewitt and Ross [2,(27.46)].

acts irreducibly, and write P_α for the projection operator carrying $H(V^{(\varphi)})$ onto $H_\alpha(\alpha\epsilon A)$. Then we have $P_\alpha V_x^{(\varphi)} = V_x^{(\varphi)} P_\alpha$ for $x \epsilon G$ and $\alpha \epsilon A$. Let $H_o = H(U)$. Then H_o is a closed linear subspace of $H(V^{(\varphi)})$ on which U acts irreducibly, and we have, for all $x \epsilon G_o$ and $\xi \epsilon H_o$

$$U_x \xi = V_x^{(\varphi)} \xi \ .$$

Since we have $I = \sum_{\alpha\epsilon A} P_\alpha$ on H, we have $H_o \subseteq \bigoplus_{\alpha\epsilon A} P_\alpha(H_o)$ and thus there id sn $\alpha_o \epsilon A$ such that $P_{\alpha_o}(H_o) \neq \{0\}$. We have

(1) $$P_{\alpha_o} U_x \xi = P_{\alpha_o} V_x^{(\varphi)} \xi = V_x^{(\varphi)} P_{\alpha_o} \xi, \quad \text{for } \xi \epsilon H_o \text{ and } x \epsilon G_o,$$

and so we see that the subspace $P_{\alpha_o}(H_o)$ of H_{α_o} is invariant under the operators $V_x^{(\varphi)}$ for $x \epsilon G_o$.

We now show that the set of operators $\{V_x^{(\varphi)} : x \epsilon G_o\}$ is irreducible on $P_{\alpha_o}(H_o)$. Assume that N is a proper subspace of $P_{\alpha_o}(H_o)$ invariant under $\{V_x^{(\varphi)} : x \epsilon G_o\}$. The set $P_{\alpha_o}^{-1}(N) \cap H_o$ is clearly a proper subspace of H_o. For $\xi \epsilon P_{\alpha_o}^{-1}(N)$ and $x \epsilon G_o$ we have

$$P_{\alpha_o} V_x^{(\varphi)} \xi = V_x^{(\varphi)} P_{\alpha_o} \xi \epsilon V_x^{(\varphi)} N \subseteq N$$

and hence we see that $P_{\alpha_o}^{-1}(N)$ is invariant under all $V_x^{(\varphi)} (x \epsilon G_o)$. This is an impossibility.

Next apply Schur's lemma to the sets of operators $\{U_x : x \epsilon G_o\}$ and $\{V_x^{(\varphi)} : x \epsilon G_o\}$ on the domains H_o and $P_{\alpha_o}(H_o)$, respectively. Since $P_{\alpha_o}(H_o) \neq \{0\}$ and since (1) holds, P_{α_o} is one-to-one on H_o, and thus is a linear isometry from H_o onto $P_{\alpha_o}(H_o)$. Equality (1) thus shows that the representation U of G_o on H_o is equivalent to the representation $x \longmapsto V_x^{(\varphi)}$ of G_o on $P_{\alpha_o}(H_o)$. The latter representation extends to the irreducible representation $x \longmapsto V_x^{(\varphi)}$ of G on H_{α_o}, and hence U likewise has a finite-dimensional irreducible extension.

Assume now that (ii) holds. Let j denote the homomorphism $x \longmapsto x$ of G_o into G. In accordance with (2.2) supra, $b(j)$ is defined so that the diagram

$$\begin{array}{ccc} G_o & \xrightarrow{\quad j \quad} & G \\ \beta_o \big\downarrow & & \big\downarrow \beta \\ bG_o & \xrightarrow{\quad b(j) \quad} & bG \end{array}$$

commutes. Let $x_o \neq e$ be an element of bG_o such that $b(j)(x_o) = e$. Since bG_o is compact, there is a finite-dimensional irreducible representation U of bG_o and an element ξ of $H(U)$ such that $U_{x_o}\xi \neq \xi$. Since the map $f : x \longmapsto U_x\xi$ of bG_o into $H(U)$ is continuous, the set

$$f^{-1}(N) = \{x \in bG_o : U_x\xi \in N\}$$

is open in bG_o, for every open subset N of $H(U)$. Let N be an open neighborhood of 0 in H such that $\xi + N$ and $U_{x_o}\xi + N$ are disjoint. Then the sets $V_1 = f^{-1}(U_{x_o}\xi + N)$ and $V_2 = f^{-1}(\xi + N)$ are disjoint open neighborhoods in bG_o of x_o and e, respectively. Since $\beta_o(G_o)$ is a dense subgroup of bG_o, for each neighborhood A of x_o in bG_o there is an element x_A in the set $\beta_o(G_o) \cap A \cap V_1$.

Now, the representation $V^{(o)} : y \longmapsto U_{\beta_o(y)}$ of G_o is finite-dimensional. In fact, $V^{(o)}$ acts irreducibly on $H(U)$. For let M be a nontrivial subspace of $H(U)$ invariant under all $V_y^{(o)}$ $(y \in G_o)$. Then M is invariant under all $U_{\beta_o(y)}$ $(y \in G_o)$. Let $\eta \in M$, and let $\zeta \in H(U)$ be orthogonal to M. Let $x \in bG_o$, and let $\{x_\gamma\}_{\gamma \in \Gamma}$ be a net in $\beta_o(G_o)$ such that $\lim_\gamma x_\gamma = x$. Then we have

$$< U_x \eta, \zeta > = \lim_\gamma < U_{x_\gamma} \eta, \zeta > = 0,$$

since $U_{x_\gamma} \eta \in M$ for $\gamma \in \Gamma$. Since ζ was an arbitrary element orthogonal to M in $H(U)$ we see that $U_x \eta \in M$ for $x \in bG_o$. Thus M is also an invariant subspace of $H(U)$ for the representation U, and so $M = H(U)$. Hence the representation $V^{(o)}$ of G_o has an extension V to all of G which is finite-dimensional and irreducible.

For each neighborhood A of x_o in G_o let $y_A \in G_o$ be such that $\beta_o(y_A) = x_A$. Then we have

$$(2) \qquad V_{y_A}^{(o)} \xi = U_{\beta_o(y_A)} \xi = U_{x_A} \xi \in (U_{x_o} \xi + N).$$

The mapping $z \longmapsto V_z$ maps G continuously into the compact group $\mathcal{U}(H(V))$, and thus by the universal property of bG there is a representation $\widetilde{V}$ mapping bG continuously into $\mathcal{U}(H(V))$ such that $\widetilde{V}_{\beta(z)} = V_z$ for all $z \in G$. Then by (2) we see for each neighborhood A of x_o that

$$(3) \qquad V_{\beta \circ j(y_A)} \xi = V_{\beta(y_A)} \xi = V_{y_A} \xi = V_{y_A}^{(o)} \xi \in (U_{x_o} \xi + N).$$

As A follows the neighborhood filter of x_o we have $x_A \longrightarrow x_o$, and hence we obtain

$$\beta \circ j(y_A) = b(j) \circ \beta_o(y_A) = b(j)(x_A) \longrightarrow b(j)(x_o) = e.$$

It follows that $\lim_A [\widetilde{V}_{\beta\circ j(y_A)}\xi] = \xi$, and hence for some neighborhood A_o of x_o we have $\widetilde{V}_{\beta\circ j(y_{A_o})}\xi \in (\xi + N)$. Since the sets $(\xi + N)$ and $(U_{x_o}\xi + N)$ are disjoint, this contradicts relation (3). Thus no such x_o exists, and $b(j)$ is one-to-one.

Finally, suppose that (iii) holds. Let φ_o be a positive-definite continuous function on G_o that is almost periodic. We may write $\varphi_o = g \circ \beta_o$, where g is a continuous positive-definite function on bG_o. Let G_1 denote the subgroup $b(j)(bG_o)$ of bG. The function $b(j)$ is a topological isomorphism of bG_o onto G_1, and so $g \circ b(j)^{-1}$ is a positive-definite continuous function defined on the closed subgroup G_1 of the compact group bG. Let g_1 be a positive-definite function on bG that extends $g \circ b(j)^{-1}$, and set $\varphi = g_1 \circ \beta$. Then φ is positive-definite, continuous, and almost periodic on G, and for $x_o \in G$ we have

$$\begin{aligned}\varphi(x_o) &= g_1 \circ \beta \circ j(x_o)\\ &= g_1 \circ b(j) \circ \beta_o(x_o)\\ &= g \circ b(j)^{-1} \circ b(j) \circ \beta_o(x_o)\\ &= g \circ \beta_o(x_o)\\ &= \varphi_o(x_o).\end{aligned}$$

This shows that φ extends φ_o. The proof is thus complete.

(2.4) Corollary. In order that the conditions of (2.3) hold, it is necessary that $vG \cap G_o = vG_o$. Here $vG = \{x \in G : U_x = I$ for all finite-dimensional representations U of $G\}$, with a similar definition for vG_o.

Proof. That $vG_o \subseteq G_o \cap vG$ is obvious. If (2.3.ii) holds, then we have $G_o \cap vG \subseteq vG_o$.

(2.5) Corollary. In order that the conditions of (2.3) hold, it is necessary that the Bohr topology on G_o coincide with the topology induced on G_o by the Bohr topology of G.[1] This condition is also sufficient if G is maximally almost periodic.

Proof. Suppose that (2.3.iii) holds. Let U be a Bohr-open neighborhood of e in G_o. Then there is a neighborhood N of e in bG_o such that $U = \beta_o^{-1}(N)$. Since $b(j)$ is an isomorphism into bG, there is an open set $N_1 \subseteq bG$ such that $N = b(j)^{-1}(N_1)$, and thus we have $U = \beta_o^{-1}(b(j)^{-1}(N_1)) = j^{-1}(\beta^{-1}(N_1)) = G_o \cap \beta^{-1}(N_1)$. The neighborhood $\beta^{-1}(N_1)$ of e in G is Bohr-open in G. Therefore every Bohr-open set in G_o is "induced" by a Bohr-open set in G.

Conversely, consider an arbitrary Bohr-open set in G of the form $\beta^{-1}(N_1)$, with N_1 open in bG. We have plainly

$$G_o \cap \beta^{-1}(N_1) = j^{-1}(\beta^{-1}(N_1)) = \beta_o^{-1}(b(j)^{-1}(N_1))$$

and $\beta_o^{-1}(b(j)^{-1}(N_1))$ is Bohr-open in G_o. This shows that every open set in G_o "induced" by a Bohr-open set in G is Bohr-open in G_o. Therefore the condition stated is necessary.

Now let G be maximally almost periodic, and assume that the stated condition holds. The groups bG_o and bG are the respective completions of G_o and G for their Bohr topologies. Both β_o and β are 1-1. The map $\beta \circ j$ thus represents G_o with its Bohr topology homeomorphically as a subgroup of bG, and hence as a dense subgroup of $\overline{\beta(G_o)}$ (closure in bG). The group $\overline{\beta(G_o)}$ is complete, and so the continuous homomorphism $\beta_o \circ j^{-1} \circ \beta^{-1}$

[1] For a definition of the Bohr topology, see Alfsen and Holm [1].

defined on $\beta \circ j(G_o)$ and taking values in the complete group bG_o has a continuous homomorphic extension f to $\overline{\beta(G_o)}$ (Bourbaki [1,III,§3,prop.5]). For $x \in G_o$ we have

$$\begin{aligned} f \circ b(j) \circ \beta_o(x) &= f \circ (\beta \circ j(x)) \\ &= (\beta_o \circ j^{-1} \circ \beta^{-1}) \circ (\beta \circ j(x)) \\ &= \beta_o(x). \end{aligned}$$

Thus $f \circ b(j)(y) = y$ for $y \in \beta_o(G_o)$, and hence this identity extends to bG_o. The function $b(j)$ thus has a left inverse, and so is 1-1. The stated condition therefore implies condition (2.3.iii) when G is maximally almost periodic.

(2.6) Example. There is a locally compact group G which is maximally almost periodic, and which has a discrete abelian normal subgroup H such that G/H is compact, and such that H is not an extending subgroup (see (1.1) supra).

Proof. Let H be the restricted direct sum $\bigoplus_{n=1}^{\infty} Q$ of countably many copies of the additive group Q of rational numbers, endowed with the discrete topology. Let $D(2)$ be the direct product of countably many copies of the multiplicative group $\{-1,1\}$, with the product topology. With each $\varepsilon = (\varepsilon_n)_{n=1}^{\infty}$ in $D(2)$, associate the automorphism $\tau[\varepsilon]$ of H defined by $\tau[\varepsilon]((x_n)_{n=1}^{\infty}) = (\varepsilon_n x_n)_{n=1}^{\infty}$. Form the semidirect product $G = H \circledS_\tau D(2)$, i.e. G is the set $H \times D(2)$ with the product topology and the following group operation

$$(x,\varepsilon)(y,\delta) = (\tau[\varepsilon](y) + x,\ \varepsilon\delta).$$

The group G was shown by Murakami [1] to be maximally almost periodic, and to possess inequivalent left and right uniform structures.

Consider the character χ defined by

$$\chi(x,e) = \exp(i \sum_{n=1}^{\infty} x_n)$$

on $H \circledS_T \{e\}$. The definition makes sense since there are only finitely many nonzero summands. The character χ is continuous because H is discrete. Let U be a neighborhood of e in G. Then $U \cap D(2)$ is a neighborhood of e in $D(2)$, and so contains some $\delta \neq e$. That is to say, for some $k \geq 1$ we have $\delta_k = -1$. Let x_k be the element $(\delta_{kj})_{j=1}^{\infty}$ of H, δ_{kj} being the "Kronecker delta". Then we have $\chi(x_k,e) = \exp(i)$, whereas we have

$$\begin{aligned} \chi((0,\delta)(x_k,e)(0,\delta^{-1})) &= \chi(\tau[\delta](x),e) \\ &= \chi(-x,e) \\ &= \exp(-i). \end{aligned}$$

We thus have $\chi(X) \neq \chi(YXY^{-1})$ with $X = (x_k,e)$ and $Y = (0,\delta) \in U$. A result of Douady (Eymard [1,p.204]) shows that χ has no extension to G as a positive-definite continuous function.

It is interesting to note that G is unimodular, and contains the open compact abelian subgroup $\{0\} \circledS_T D(2)$.

(2.7) Theorem. Let G be a locally compact group and let G_o be a compact subgroup of G. Then G_o is an extending subgroup.

Proof. Let E denote the set of equivalence classes of irreducible representations of G_o that admit an extension to G. Let Σ denote the dual object of G_o. Since every representation of G_o is a direct sum of irreducible representations, it will suffice to show that $E = \Sigma$.

Let σ and τ be in E. Let U and V be representations from the classes σ and τ respectively. There are representations U' and V' with representation spaces H' and J' respectively, such that we have $H \subseteq H'$ and $J \subseteq J'$, H and J denoting the representation spaces of U and V respectively, and analogues of (2.4.1) hold. Since $H \otimes J$ may be regarded as a closed subspace of $H' \otimes J'$, it is clear that the representation $U' \otimes V'$ extends $U \otimes V$. Thus $\sigma \times \tau \in E$.

Let $\sigma \in E$, and let $U \in \sigma$. Let U' be an extension of U to G, and let D be a conjugation operator on H(U). Choose any conjugation operator $D^{\perp}$ on $H(U)^{\perp}$, the orthogonal complement of H(U) in H(U'). Set $D' = D \oplus D^{\perp}$. Then $x \longmapsto D'U'_xD'$ is a representation of G with representation space H(U'), and for $x \in G_o$ and $\xi \in H(U)$ we have

$$D'U'_xD'\xi = D'U'_xD\xi = D'U_xD\xi = DU_xD\xi.$$

Therefore the representation $x \longmapsto D'U'_xD'$ extends the representation $x \longmapsto DU_xD$. Thus we see that $\bar{\sigma} \in E$.

Let $x_o \in G_o \setminus \{e\}$. There is an irreducible representation U of G such that $U_{x_o} \neq I$. Restricting $x \longmapsto U_x$ to G_o, we have a decomposition $U = \bigoplus_{\iota \in I} U^{(\iota)}$ of U into irreducible representations of G_o on subspaces of H(U). Since $U_{x_o} \neq I$, we have $U^{(\iota)}_{x_o} \neq I$ for some $\iota \in I$. This proves that E separates points of G_o.

Combining the facts above with Hewitt and Ross [2,(27.39)], we find that $E = \Sigma$.

(2.8) Example. It can happen that a locally compact group G has an extending subgroup G_o which does not enjoy property (2.3.i). In fact, if

$G = SL(2,\mathbb{C})$ with its usual topology and G_o is the subgroup $\left\{\begin{bmatrix}1 & 0\\ 0 & 1\end{bmatrix}, \begin{bmatrix}-1 & 0\\ 0 & -1\end{bmatrix}\right\}$ of G, then G_o is a compact subgroup and thus is an extending subgroup, by theorem (2.7). On the other hand, G has no nontrivial finite-dimensional representations (Hewitt and Ross [1,(22.22.h)]) and hence by (2.1) the non-trivial character of G_o has no almost periodic positive-definite extension to G.

§3. MISCELLANEOUS EXTENSION THEOREMS

Any title more specific than the one given this section would be misleading. Here are collected results that defy all attempts at classification into a reasonable number of categories. Thus we shall introduce theorems and groups of theorems as they come.

We begin with some results that will be useful in the sequel. The first lemma is well known, and we omit the proof.

(3.1) Lemma. Let G_o be a locally compact group. Let $g_i \in \underline{\underline{P}}(G)$, $i = 1,2,3$, and suppose $g_1 = g_2 + g_3$. Let $V^{(g_1)}$ be the representation of G_o corresponding to g_1 with representation space $H(V^{(g_1)})$ and cyclic vector ζ. Then there is a positive-definite operator B on $H(V^{(g_1)})$ commuting with all operators $V_x^{(g_1)}$ $(x \in G_o)$, such that for all $x \in G_o$ we have

$$g_2(x) = \langle BV_x^{(g_1)}\zeta, \zeta\rangle .$$

(3.2) Theorem. Let G be a locally compact group, of which G_o is a closed subgroup. If f, g, and $f - g \in \underline{\underline{P}}(G_o)$ and if f has an extension in $\underline{\underline{P}}(G)$, then g has an extension in $\underline{\underline{P}}(G)$.

Proof. According to (1.5), the canonical cyclic representation $V^{(f)}$ has an extension to G in the sense of (1.4). That is to say, the representation space $H^{(f)}$ of $V^{(f)}$ is contained in a Hilbert space H' on which acts a representation V' of G, and for $x \in G_o$ and $\xi \in H^{(f)}$ we have

$$V_x^{(f)}\xi = V_x'\xi.$$

Thus for $x \in G_o$ we have

$$f(x) = \langle V_x'\zeta, \zeta \rangle .$$

Since $f = g + (f - g)$ on G_o, we may invoke the previous lemma to deduce the existence of a positive-definite operator B on $H^{(f)}$, commuting with all $V_x^{(f)}$ $(x \in G_o)$ such that

$$g(x) = \langle BV_x'\zeta, \zeta \rangle$$

for $x \in G_o$. Let A be a positive-definite square root of B. The operator A commutes with all $V_x^{(f)}$ $(x \in G_o)$ and we have

$$g(x) = \langle AV_x'A\zeta, \zeta \rangle = \langle V_x'A\zeta, A\zeta \rangle$$

for $x \in G_o$. The function $x \longmapsto \langle V_x'A\zeta, A\zeta \rangle$ defined on G is in $\underline{\underline{\underline{P}}}(G)$, and extends g.

(3.3) <u>Theorem</u>. Let G be a locally compact group, let G_o be a closed subgroup of G, and let $\varphi_o \in \underline{\underline{\underline{P}}}(G_o)$. If φ_o has an extension in $\underline{\underline{\underline{K}}}(G)$, then φ_o has an extension in $\underline{\underline{\underline{P}}}(G)$.

<u>Proof</u>. First let φ_o be an arbitrary Hermitian function in $\underline{\underline{\underline{K}}}(G_o)$ with an extension ψ in $\underline{\underline{\underline{K}}}(G)$. Write $\psi = \psi^+ + i\psi^-$, where ψ^+ and ψ^- are continuous Hermitian functions in $\underline{\underline{\underline{K}}}(G)$. Then we have for $x \in G_o$

$$\varphi_o(x) = \psi^+(x) + i\psi^-(x)$$

in which φ_o and ψ^+ are Hermitian on G_o, and hence we see that $i\psi^-$ is Hermitian on G_o. Thus for $x \in G_o$ we have

$$i\psi^-(x) = \overline{i\psi^-(x^{-1})} = -\overline{i\psi^-(x^{-1})} = -i\psi^-(x),$$

since ψ^- is Hermitian, and thus $\psi^-(x) = 0 (x \in G_o)$. Hence the Hermitian function ψ^+ is itself an extension of φ_o.

Now suppose in addition that φ_o is positive-definite. Since $\psi^+ \in \underline{\underline{\underline{K}}}(G)$, we may write $\psi^+ = \psi_1 - \psi_2$, where ψ_1 and ψ_2 are in $\underline{\underline{\underline{P}}}(G)$, and so obtain

$$\psi_1(x) = \psi_o(x) + \psi_2(x)$$

for $x \in G_o$. Applying (3.2) with $f = \psi_1$ and $g = \varphi_o$ we see that φ_o has an extension in $\underline{\underline{\underline{P}}}(G)$.

The next group of results, comprising (3.4) through (3.7), is concerned with the class of all locally compact groups with the extension property.

(3.4) <u>Discussion and example</u>. That every closed subgroup of a group with the extension property has the extension property, is obvious. The following example shows that a locally compact group continuously represented in a group with the extension property need <u>not</u> have the extension property, even when the representing homomorphism is injective.

Let G be as in (2.6), and let bG denote its Bohr compactification. Since G is maximally almost periodic, the canonical map $\beta : G \longrightarrow bG$ is 1-1 and continuous. The group bG, being compact, has the extension property. The group G, as noted in (2.6), does not.

(3.5) <u>Theorem</u>. Let G be a locally compact group with the extension property, and let H be a closed normal subgroup of C. Then G/H has the extension property.

Proof. Let G_o be a closed subgroup of G/H, and let σ denote the natural quotient homomorphism from G onto G/H. Set $\widetilde{G}_o = \sigma^{-1}(G_o)$. Suppose that φ_o is a positive-definite continuous function on G_o, and set $\widetilde{\varphi}_o = \varphi_o \circ \sigma$, so that $\widetilde{\varphi}_o \in \underline{\underline{P}}(G_o)$ and $\widetilde{\varphi}_o$ is constant on cosets of H. Let $\widetilde{\varphi}$ be an extension of $\widetilde{\varphi}_o$ in $\underline{\underline{P}}(G)$. Since $G_1 = \{y \in G : \widetilde{\varphi}(y) = \widetilde{\varphi}(e)\}$ is a closed subgroup of G on whose cosets $\widetilde{\varphi}$ is constant, and since G_1 clearly contains H, the function $\widetilde{\varphi}$ is constant on cosets of H. Therefore we have $\widetilde{\varphi} = \widetilde{\varphi} \circ \sigma$ for some $\varphi \in P(G/H)$. If $x \in \widetilde{G}_o$, then we have

$$\varphi \circ \sigma(x) = \widetilde{\varphi}(x) = \widetilde{\varphi}_o(x) = \varphi_o \circ \sigma(x),$$

and thus φ extends φ_o.

(3.6) Example. Let G be any group without the extension property. Since G is a continuous homomorphic image of the discrete group G_d, the extension property is not preserved by continuous homomorphic images.

(3.7) Theorem. Let G be a locally compact group, and let H be a closed subgroup of G of finite index such that H has the extension property. Then G has the extension property.

Proof. Let G_o be a closed subgroup of G, let $\varphi_o \in \underline{\underline{P}}(G_o)$, and let V be a representation of G_o on $H(V)$ such that we have $\varphi_o(x) = \langle V_x\zeta, \zeta \rangle$ for $x \in G_o$, for a suitable cyclic vector ζ. Let $\widetilde{V}$ be an extension of the representation $x \longmapsto V_x$, restricted to the group $G_o \cap H$, to the group H. Let X be a (finite) complete set of left coset representatives for H in G_oH, chosen so that $e \in X \subseteq G_o$. Then X is also a complete set of left coset representatives for $G_o \cap H$ in G_o.

Define φ' on G_oH by setting

$$(1) \qquad \varphi'(z) = \varphi'(xy) = < \tilde{V}_y \zeta, V_{x^{-1}} \zeta > ,$$

where we have written $z = xy$ with $x \in X$ and $y \in H$. Define $\varphi'(z) = 0$ for $z \notin G_o H$. Further define

$$(2) \qquad {}^x f : y \longmapsto < \tilde{V}_y \zeta, V_{x^{-1}} \zeta >$$

for $x \in X$, $y \in H$, so that for each $x \in X$ we have ${}^x f \in \underline{\underline{K}}(H)$. Since H is open in G, setting

$${}^x f(y) = 0 \qquad y \notin H, x \in X$$

extends ${}^x f$ to G as an element of $\underline{\underline{K}}(G)$. Relations (1) and (2) above show that for $x \in X$ we have

$$\varphi' \xi_{(xH)} = {}_{x^{-1}}({}^x f),$$

and hence that $\varphi' \xi_{(xH)} \in \underline{\underline{K}}(G)$ for each $x \in X$. Since the relation

$$\varphi' = \sum_{x \in X} \varphi' \xi_{(xH)}$$

holds on G, we have $\varphi' \in \underline{\underline{K}}(G)$. For $z \in G_o$ we have $z = xy$ with $x \in X$, $y \in G_o \cap H$, so that we also have

$$\begin{aligned} \varphi'(z) &= < V_y \zeta, V_{x^{-1}} \zeta > \\ &= < V_x V_y \zeta, \zeta > \\ &= < V_z \zeta, \zeta > \\ &= \varphi_o(z). \end{aligned}$$

This shows that φ' extends φ_o as an element of $\underline{\underline{K}}(G)$. Theorem (3.3) now shows that the claim is valid.

(3.8) Notation and discussion. Let G be a topological group, and let G_o be a closed subgroup of G. We shall write $\underline{\underline{P}}(G_o,G)$ for the set $\{\varphi|G_o : \varphi \in \underline{\underline{P}}(G)\}$, and $\underline{\underline{K}}(G_o,G)$ for the set $\{\varphi|G_o : \varphi \in \underline{\underline{K}}(G)\}$. We shall also write $\underline{\underline{K}}_h(G_o)$ and $\underline{\underline{K}}_h(G_o,G)$ respectively for the sets of Hermitian elements of $\underline{\underline{K}}(G_o)$ and $\underline{\underline{K}}(G_o,G)$.

Theorem (3.3) can now be stated as

(1) $$\underline{\underline{P}}(G_o,G) = \underline{\underline{P}}(G_o) \cap \underline{\underline{K}}(G_o,G).$$

The first paragraph of the proof of theorem (3.3) shows that we also have

(2) $$\underline{\underline{K}}_h(G_o,G) = \underline{\underline{K}}_h(G_o) \cap \underline{\underline{K}}(G_o,G).$$

These definitions allow us to draw attention to the following obvious theorems.

(3.9) Theorem. Let G be a locally compact group, and let G_o be a closed subgroup of G. If $\varphi_1,\varphi_2 \in \underline{\underline{P}}(G_o,G)$, then $\bar{\varphi}_1,\varphi_1^{\star}$, and $\varphi_1\varphi_2$ belong to $\underline{\underline{P}}(G_o,G)$. Also $\alpha_1\varphi_1 + \alpha_2\varphi_2$ is in $\underline{\underline{P}}(G_o,G)$ for all $\alpha_1,\alpha_2 \geq 0$.

The set $\underline{\underline{K}}(G_o,G)$ is the linear span of $\underline{\underline{P}}(G_o,G)$, and is a self-conjugate subalgebra of $\underline{\underline{K}}(G_o)$, invariant under translation.

(3.10) Theorem. Let G and G_o be as in (3.9). The following conditions are equivalent.

(i) The subgroup G_o is extending.

(ii) We have $\underline{\underline{P}}(G_o,G) = \underline{\underline{P}}(G_o)$.

(iii) We have $\underline{\underline{K}}(G_o,G) = \underline{\underline{K}}(G_o)$.

(iv) The algebra $\underline{\underline{K}}(G_o,G)$ is closed in $\underline{\underline{K}}(G_o)$ for the topology of uniform convergence on compact sets.

(3.11) Remark. The conditions in theorem (3.10) evidently imply the following.

(v) The set $\underline{\underline{P}}(G_o,G)$ is closed in $\underline{\underline{P}}(G_o)$ for the topology of compact convergence.

However, equivalence of condition (v) to the conditions of theorem (3.10) has not been established, save in a particular case. See (4.23) *infra*.

We conclude this section with two examples concerning ascending chains of subgroups.

(3.12) Example. Let G be the group of matrices of the form $(x,t) = \begin{bmatrix} e^t & x \\ 0 & 1 \end{bmatrix}$ with x and t arbitrary real numbers and with the topology of G induced by the usual topology for $GL(2,\mathbb{R})$. Let $G_o = \{(x,0) : x \in \mathbb{R}\}$. Then G is the well-known "ax + b" group, and G_o is the "pure translation subgroup". Douady has shown (see Hewitt and Ross [2,(34.48.e)][1] that G_o has a continuous character χ with no extension in $\underline{\underline{P}}(G)$. Thus (G,G_o,χ) is not an extension triple.

Now, for $n = 1,2,\ldots,$ let $G_n = \{(x, \frac{k}{2^n}) : k \in \mathbb{Z}, x \in \mathbb{R}\}$. For $n > 0$, the group G_o is closed and normal in G_n, being closed and normal in G. The group G_n/G_o is plainly topologically isomorphic with the additive group $\{\frac{k}{2^n} : k \in \mathbb{Z}\}$, and so G_o is open in G_n. Therefore, for $n = 1,2,\ldots,$ (G_n,G_o,φ_o) is an extension triple for each $\varphi_o \in \underline{\underline{P}}(G_o)$.

The group $\bigcup_{n=1}^{\infty} G_n$ is dense in G, and so the character χ can have no extension in $P(\bigcup_{n=1}^{\infty} G_n)$, for then it would have an extension in $\underline{\underline{P}}(G)$ (the argument here is analogous to that used at the beginning of the proof of theorem (2.1), so is omitted).

[1] There is, however, a misprint in the matrices used in *loc. cit.*

Thus it is possible to find subgroups $G_o, G_1, G_2, \ldots$ of a group G and a function $\varphi_o \in \underline{\underline{P}}(G_o)$ such that (G_n, G_o, φ_o) is an extension triple, for for $n = 0,1,2,\ldots,$ while $(\bigcup_{n=1}^{\infty} G_n, G_o, \varphi_o)$ is not.

(3.13) Example. The union of an ascending chain of extending subgroups need not be an extending subgroup.

Proof. Let $\mathbb{Z}_3$ denote the additive group of integers modulo 3. The elements of $\mathbb{Z}_3$ we write as $\underline{0}, \underline{1}, \underline{2}$. Let H be the discretely topologized direct sum $\bigoplus_{n=1}^{\infty} \mathbb{Z}_3$, and let $D(2)$ be the direct product of countably many copies of the multiplicative group $\{-1,1\}$, with the product topology. With each $\varepsilon = (\varepsilon_n)_{n=1}^{\infty}$ in $D(2)$, with $\varepsilon_n \in \{-1,1\}$ for $n = 1,2,\ldots,$ associate the automorphism $\tau[\varepsilon]$ of H defined by the rule $\tau[\varepsilon]((x_n)) = (\varepsilon_n x_n)$, where each product $\varepsilon_n x_n$ is to be interpreted as integral multiplication modulo 3. Form the semidirect product $G = H \circledS_\tau D(2)$. Both H and $D(2)$ may be regarded as subgroups of G in the natural way; H is a discrete closed normal subgroup, while $D(2)$ is a compact open subgroup.

Consider the character χ defined on H by setting

$$\chi(x,e) = \exp\left(\frac{2\pi i}{3} \sum_{n=1}^{\infty} x'_n\right),$$

where the (finite) sum $\sum_{n=1}^{\infty} x'_n$ is formed by choosing, for each n, an integer x'_n congruent to x_n modulo 3: the value of $\exp\left(\frac{2\pi i}{3} \sum_{n=1}^{\infty} x'_n\right)$ is independent of the representatives x'_n chosen. Since H is discrete, the character χ is continuous on H. Let U be any neighborhood of e in G. Then $U \cap D(2)$ is a neighborhood of e in $D(2)$, and so contains an element $(0,\delta)$ with $\delta \neq e$. That is to say, for some $k \geq 1$ we have $\delta_k = -1$. Let x_k be the element of H whose j^{th} component is $\underline{1}$ if $j = k$, and $\underline{0}$ otherwise. Then we have $\chi(x_k, e) = \exp\left(\frac{2\pi i}{3}\right)$, whereas we have

$$\chi((0,\delta)(x_k,e)(0,\delta)^{-1}) = \chi(\tau[\delta](x),e)$$
$$= \chi(-x,e)$$
$$= \exp\left(-\frac{2\pi i}{3}\right).$$

We thus have $\chi(X) \neq \chi(YXY^{-1})$ with $X = (x_k,e)$ and $Y = (0,\delta) \in U$. A result of Douady (see Hewitt and Ross loc. cit.) shows that χ has no extension in $\underline{\underline{P}}(G)$. Therefore H is not an extending subgroup of G.

Let $H_k = \{(x_n) \in H : x_j = 0 \text{ for } j > k\}$. The set H_k is actually a closed subgroup of H, and, when identified with the subgroup $H_k \circledS_\tau D(2)$ of G, forms a closed subgroup of G. Since H_k is finite, it is compact, and so by theorem (2.7) each H_k is an extending subgroup. The subgroup $\bigcup_{k=1}^{\infty} H_k$ is clearly H, which is not an extending subgroup.

§4. EXTENSION OF ELEMENTS OF THE FOURIER ALGEBRA

In the previous section we introduced the algebra $\underline{\underline{K}}(G_o,G)$ for a closed subgroup G_o of a locally compact group G. In this section we shall prove that $\underline{\underline{K}}(G_o,G)$ always contains the "algèbre de Fourier" of G_o discussed in Eymard [1]. This result was stated without proof in Eymard [2], and no proof is in print to the knowledge of the present writer.[1] Our proof culminates with (4.21) *infra*.

(4.1) Definition. Let G be a locally compact group.

(i) The set of functions $f * \overline{g^\star}$, with f and g in $L^2(G)$, will be called the Fourier algebra (Eymard: "algèbre de Fourier") of G, and will be denoted by $A(G)$. Eymard [1] has shown that $A(G)$ is a closed ideal in $\underline{\underline{K}}(G)$.[2]

[1] I have since discovered that a proof has been published by C. S. Herz.

[2] Henceforth $\underline{\underline{K}}(G)$ will be endowed with the norm $\|\ \|$ of Eymard loc. cit.

(ii) If $1 < p < \infty$, let $p' = \frac{p}{p-1}$. The set of functions of the form $\sum_{k=1}^{\infty} f*\overline{g_k^{\star}}$, with $f_k \in L^p(G)$ and $g_k \in L^{p'}(G)$ for $k = 1,2,\ldots$ and with $\sum_{k=1}^{\infty} \| f_k \|_p \| g_k \|_{p'} < \infty$ will be denoted by $A_p(G)$ (see (4.3) <u>infra</u>).

(4.2) <u>Proposition</u>. Let G be a locally compact group. Then we have $A_2(G) = A(G)$.

<u>Proof</u>. It is clear from the definitions that $A(G) \subseteq A_2(G)$. Conversely, suppose that $f \in A_2(G)$, so that $f = \sum_{k=1}^{\infty} g_k*\overline{h_k^{\star}}$ with $\{g_k\}_{k=1}^{\infty} \subseteq L^2(G)$, $\{h_k\}_{k=1}^{\infty} \subseteq L^2(G)$, and $\sum_{k=1}^{\infty} \| g_k \|_2 \| h_k \|_2 < \infty$.

Since $A(G)$ is an algebra, we have $f_n = \sum_{k=1}^{n} g_k*\overline{h_k^{\star}} \in A(G)$ for each n. Let $\varepsilon > 0$, and let n_o be a positive integer such that for $m > n > n_o$ we have $\sum_{k=n+1}^{m} \| g_k \|_2 \| h_k \|_2 < \varepsilon$. Then by Eymard [1,(2.15.1°) and (3.1)] we have for such m and n

$$\begin{aligned} \| f_m - f_n \|_u &\leq \| f_m - f_n \| \\ &\leq \sum_{k=n+1}^{m} \| g_k * h_k \| \\ &\leq \sum_{k=n+1}^{m} \| g_k \|_2 \| h_k \|_2 \\ &< \varepsilon. \end{aligned}$$

Thus the sequence $\{f_n\}_{n=1}^{\infty}$ is Cauchy in $\underline{\underline{K}}(G)$ and in $C(G)$, and thus there is a function $f' \in K(G)$ such that $f_n \longrightarrow f'$ uniformly on G; but then for all $n \geq 1$ we have

$$\begin{aligned} \| f_n - f \|_u &= \| \sum_{k=n+1}^{\infty} g_k * \overline{h_k^{\star}} \|_u \\ &\leq \sum_{k=n+1}^{\infty} \| g_k * \overline{h_k^{\star}} \|_u \\ &\leq \sum_{k=n+1}^{\infty} \| g_k \|_2 \| h_k \|_2 \end{aligned}$$

by Hewitt and Ross [1,(20.16)], and so $f_n \longrightarrow f$ uniformly on G. It follows that $f = f'$, so that $f \in \underline{\underline{\underline{K}}}(G)$, and since A(G) is closed in $\underline{\underline{\underline{K}}}(G)$ and $f_n \longrightarrow f$ in $\underline{\underline{\underline{K}}}(G)$, we must have $f \in A(G)$. We have thus proved that $A_2(G) \subseteq A(G)$.

(4.3) Theorem. Let G be a locally compact group. Then for $1 < p < \infty$ we have $A_p(G) \subseteq C_o(G)$.

Proof. Let $p' = \frac{p}{p-1}$ and let $g \in L^p(G)$ and $h \in L^{p'}(G)$. Then by Hewitt and Ross [1,(20.16)] and its proof we have that $g * \overline{h^*}$ is defined, since $(\overline{h^*})^* = \bar{h} \in L^{p'}(G)$, and is a function in $C_o(G)$ such that

$$\begin{aligned} \| g * \overline{h^*} \|_u &\leq \| g\|_p \| (\overline{h^*})^* \|_{p'} \\ &= \| g\|_p \| \bar{h}\|_{p'} \\ &= \| g\|_p \| h\|_{p'} \,. \end{aligned}$$

Therefore if $\{g_k\}_{k=1}^{\infty}$ and $\{f_k\}_{k=1}^{\infty}$ are sequences in $L^p(G)$ and $L^{p'}(G)$ respectively such that

$$\sum_{k=1}^{\infty} \| g_k\|_p \| h_k\|_p < \infty$$

then the series

$$\sum_{k=1}^{\infty} g_k * \overline{h_k^*}$$

is absolutely convergent in the Banach space $C_o(G)$, and thus represents a function in $C_o(G)$. Thus $A_p(G) \subseteq C_o(G)$.

(4.4) Theorem. Let G be a locally compact group. If f is a function in A(G), then there are functions $f_j \in L^2(G)$, $j \in \{0,1,2,3\}$ such that

$$(*) \qquad f = \sum_{j=0}^{3} i^j f_j * \overline{f_j^\star} .$$

If f is an Hermitian function in $A(G)$, then we can take $f_j = 0$ for $j = 1,3$, so that we have

$$f = f_o * \overline{f_o^\star} - f_2 * \overline{f_2^\star} .$$

If f is a positive-definite function on G, the following statements are equivalent.

(i) The cyclic representation $V^{(f)}$ is equivalent to a subrepresentation of the left regular representation of G.

(ii) There exists a function $f \in L^2(G)$ such that $f = f * \overline{f^\star}$.

(iii) The function f is in $A(G)$.

Proof. Let f be a function in $A(G)$. Then there are functions $g,h \in L^2(G)$ such that $f = g * \overline{h^\star}$. The identity

$$f = \sum_{j=0}^{3} i^j \left(\frac{g+i^j h}{2}\right) * \overline{\left(\frac{g+i^j h}{2}\right)^\star} ,$$

easily verified and analogous to the well-known "polar identity," proves (*) with $f_j = \frac{g+i^j h}{2}$, $j \in \{0,1,2,3\}$.

Now we note that for each j, $f_j * \overline{f_j^\star}$ is positive-definite (see, for example, Hewitt and Ross [2,(32.43.e)]) and therefore we see that the functions

$$g_o = f_o * \overline{f_o^\star} - f_2 * \overline{f_2^\star}$$

and

$$g_1 = f_1 * \overline{f_1^\star} - f_3 * \overline{f_3^\star}$$

are Hermitian on G. Thus we have for $x \in G$

$$f(x^{-1}) = g_o(x^{-1}) + ig_1(x^{-1}) = \overline{g_o(x)} + i\overline{g_1(x)},$$

whereas for all $x \in G$ we also have

$$\overline{f(x)} = \overline{g_o(x) + ig_1(x)} = \overline{g_o(x)} - \overline{ig_1(x)} .$$

If f is Hermitian, so that $f(x^{-1}) = \overline{f(x)}$ for all $x \in G$, we therefore have $g_1(x) = 0$ for $x \in G$, so that we may take $f_1 = f_3 = 0$ in (*), giving

$$f = g_o = f_o * \overline{f_o^{\star}} - f_2 * \overline{f_2^{\star}} .$$

The equivalence of (i), (ii) and (iii) will now be proved by the schema (i) $\Longrightarrow$ (ii) $\Longrightarrow$ (iii); (ii) $\Longrightarrow$ (i); (iii) $\Longrightarrow$ (i). Throughout we shall suppose that f is positive-definite.

If (i) holds, then there is a cyclic vector $f' \in L^2(G)$ such that for $x \in G$ we have

$$f(x) = \langle R_x f', f' \rangle$$

where R denotes the left regular representation of G on the Hilbert space $L^2(G)$. Thus we have

$$\begin{aligned} f(x) &= \langle R_x f', f' \rangle \\ &= \int \left(_{x^{-1}} f'\right) (y)\overline{f'}(y)d\lambda(y) \\ &= \int f'(x^{-1}y)\overline{f'}(y)d\lambda(y) \\ &= \int f'(y)\overline{f'}(xy)d\lambda(y) \\ &= \int \overline{(\overline{f'})^{\star}} (y^{-1})\overline{f'}(xy)d\lambda(y) \\ &= \overline{f'} * \overline{(\overline{f'})^{\star}} (x), \end{aligned}$$

where we have used left invariance of λ and the convolution formula (20.16.i) of Hewitt and Ross [1]. Therefore (ii) holds with $f_o = \overline{f'}$.

The fact that (ii) implies (iii) is a trivial consequence of (4.1.i) <u>supra</u>.

We now prove that (ii) implies (i), and so suppose again that (ii) holds. Then we have for $x \in G$

$$f(x) = f * \overline{f^*}(x) = < R_x\overline{f}, \overline{f} > ,$$

as before. This implies, by Hewitt and Ross [2,(32.8.b)] that the representation R, with all operators R_x restricted to the closure of the subspace of $L^2(G)$ spanned by the set $\{R_x(\overline{f_o}) : x \in G\}$ is equivalent to $V^{(f)}$. Thus property (i) holds.

Finally, we prove that (iii) implies (i), and so suppose that (iii) holds. Since f is certainly Hermitian we may as above write

$$f = f_o * \overline{f_o^*} - f_2 * \overline{f_2^*}$$

with $f_o, f_2 \in L^2(G)$. Since f, $f_o * \overline{f_o^*}$, and $f_2 * \overline{f_2^*}$ are all positive-definite, we may apply lemma (3.1) with $g_1 = f_o * \overline{f_o^*}$, $g_2 = f$ and $g_3 = f_2 * \overline{f_2^*}$ to see that there is a positive-definite operator B on $H(V^{(f_o*\overline{f_o^*})})$ commuting with all $V_x^{(f_o*\overline{f_o^*})}$ for $x \in G$, such that we have

$$f(x) = < BV_x^{(f_o*\overline{f_o^*})} \zeta, \zeta >$$

for $x \in G$, ζ being a cyclic vector for $V^{(f_o*\overline{f_o^*})}$. Let A be a positive-definite square root of B. The operator A commutes with all $V_x^{(f_o*\overline{f_o^*})}$ $(x \in G)$, and we have

$$f(x) = < V_x^{(f_o*\overline{f_o^*})} A\zeta, A\zeta > .$$

Invoking Hewitt and Ross [32.8.b) once more, we see that $V^{(f)}$ is equivalent to the representation $V^{(f_o * \overline{f_o^*})}$ with all operators $V_x^{(f_o * \overline{f_o^*})}$ restricted to the closure of the linear span of the set $\{V_x^{(f_o * \overline{f_o^*})} A\zeta : x \in G\}$. Let T be an intertwining isometry from $H(V^{(f)})$ onto this subspace for $V^{(f)}$ and $V^{(f_o * \overline{f_o^*})}$. We have for all $x \in G$

$$TV_x^{(f)} = V_x^{(f_o * \overline{f_o^*})} T.$$

Now we apply the proven implication (ii) $\Longrightarrow$ (i) to $V^{(f_o * \overline{f_o^*})}$. That is, there exists a linear isometry S from $H(V^{(f_o * \overline{f_o^*})})$ into $L^2(G)$ such that for all $x \in G$ we have

$$SV_x^{(f_o * \overline{f_o^*})} = R_x S.$$

It follows that ST is a linear isometry from $H(V^{(f)})$ into $L^2(G)$ for which we have for all $x \in G$

$$\begin{aligned}(ST)V_x^{(f)} &= S(V^{(f_o * \overline{f_o^*})} T) \\ &= (R_x S)T \\ &= R_x(ST),\end{aligned}$$

proving that $V^{(f)}$ is equivalent to a subrepresentation of R. This completes the proof.

(4.5) Lemma. Let G be a locally compact noncompact group, and let $f \in C_o(G)$. Then for each $\varepsilon > 0$ there is a sequence $\{x_1, \ldots, x_m\}$ in G and positive real numbers $\alpha_1, \alpha_2, \ldots, \alpha_m$ such that $\sum_{j=1}^{m} \alpha_j = 1$ and

$$\left\| \sum_{j=1}^{m} \alpha_j ({}_{x_j} f) \right\|_u < \varepsilon.$$

Proof. Let N be a compact subset of G, and suppose that we have $x \in G$ and $xN \cap N \neq \emptyset$. Let $y \in xN \cap N$. Then we have $y \in N$ and $y = xz$ for some $z \in N$, so that $x = yz^{-1} \in NN^{-1}$. Thus the set $\{x \in G : xN \cap N \neq \emptyset\}$ is contained in the compact set NN^{-1}. Hence for some $x \in G$ we must have $xN \cap N = \emptyset$.

Let $\varepsilon > 0$ be given and let $N_1 = \{x \in G : |f(x)| \geq \frac{\varepsilon}{2}\}$.

Inductively, one may continue and obtain a sequence $\{x_1, x_2, \ldots, x_m\}$, where m is a positive integer with $m > \frac{2\|f\|_u}{\varepsilon}$, such that the x_jN_1 are pairwise disjoint.

Let $x \in G$. If for some $j_o \in \{1,2,\ldots,m\}$ we have $x \in x_{j_o}N_1$, then we have $\left|\left({}_{x_{j_o}^{-1}}f\right)(x)\right| = \left|f\left(x_{j_o}^{-1}x\right)\right| \leq \|f\|_u$. For $j' \neq j_o$ we have $x \notin x_{j'}N_1$, whence $x_{j'}^{-1}x \notin N_1$, and so we have $\left|\left({}_{x_{j'}^{-1}}f\right)(x)\right| = \left|f\left(x_{j'}^{-1}x\right)\right| < \frac{\varepsilon}{2}$.

Therefore we have

$$\begin{aligned}
\left|\sum_{j=1}^{m} \frac{1}{m}\left({}_{x_j^{-1}}f\right)(x)\right| &\leq \frac{1}{m}\sum_{j=1}^{m}\left|\left({}_{x_j^{-1}}f\right)(x)\right| \\
&= \frac{1}{m}\left(\left|\left({}_{x_{j_o}^{-1}}f\right)(x)\right| + \sum_{\substack{j'=1 \\ j'\neq j_o}}^{m}\left|\left({}_{x_{j'}^{-1}}f\right)(x)\right|\right) \\
&\leq \frac{\|f\|_u}{m} + \frac{m-1}{m}\cdot\frac{\varepsilon}{2} \\
&< \frac{\varepsilon}{2} + \frac{\varepsilon}{2} \\
&= \varepsilon.
\end{aligned}$$

If on the other hand we have, for all j, $x \notin x_jN$ then we have as above for $j = 1,2,\ldots,m$ that $\left|\left({}_{x_j^{-1}}f\right)(x)\right| < \frac{\varepsilon}{2}$, and so

$$\left| \sum_{j=1}^{m} \frac{1}{m}\left({}_{x_j^{-1}}f \right)(x) \right| < \frac{\varepsilon}{2} .$$

Thus we see that

$$\left\| \sum_{j=1}^{m} \frac{1}{m}\left({}_{x_j^{-1}}f \right) \right\|_u < \varepsilon .$$

This proves the claim.

(4.6) Corollary. Let G be a locally compact noncompact group, and let φ be a positive-definite function in $A(G)$. Then the representation $V^{(\varphi)}$ has no nontrivial finite-dimensional subrepresentations.

Proof. Since $\varphi \in A(G)$, we have by (4.3) $\varphi \in C_o(G)$, and hence $|\varphi|^2 \in C_o(G)$. Thus for each $\varepsilon > 0$ we see by lemma (4.5) that there is a sequence $\{x_1,\ldots,x_m\}$ in G and positive real numbers $\alpha_1,\alpha_2,\ldots,\alpha_m$ such that $\sum_{j=1}^{m} \alpha_j = 1$ and

$$\left\| \sum_{j=1}^{m} \alpha_j \left({}_{x_j}f \right) \right\|_u < \varepsilon ,$$

where we have written f for $|\varphi|^2$. Godement [1,Théorème 16(3)] has shown that the assertion that φ enjoys this property is equivalent to the assertion that $V^{(\varphi)}$ contain no nontrivial finite-dimensional subrepresentations.

Remark. Let G be a locally compact noncompact group. Then every finite-dimensional subrepresentation of the regular representation of G, if non-trivial, contains a nontrivial cyclic subrepresentation, which must be equivalent to $V^{(f)}$, for some positive-definite continuous function f. Theorem (4.4) asserts that this implies that $f \in A(G)$, which is absurd in view of (4.6). Thus we see that the regular representation of G contains no nontrivial finite-dimensional subrepresentations. This statement should be compared with Mackey [1,Theorem (3.2),Cor.], which makes the same assertion only for groups G with a countable base for all open sets. The present result is thus slightly more general, and possesses a strikingly more elementary proof.

The next corollary to (4.3) shows that the results of this section are disjoint from the considerations of §2.

(4.7) Corollary. Let G be a locally compact noncompact group, and let φ be a positive-definite function in $A(G)$. If φ is also almost periodic, then $\varphi = 0$.

Proof. This is immediate on combining (2.1) and (4.4).

Having completed the preliminaries, we now proceed to the main result of this section. Several lemmas are needed.

(4.8) Mackey's Lemma. Let G be a locally compact group with a countable base for its topology, and let G_o be a closed subgroup of G. Then there is a Borel set B in G such that

(i) B intersects each right coset of G_o in exactly one point,

(ii) $(G_oK) \cap B$ has compact closure for each compact $K \subseteq G$, and

(iii) $G_o \cap B = \{e\}$.

Proof. See Mackey [1] for existence of B with properties (i), (ii). Let $b \in G_o \cap B$. Then $B^* = b^{-1}B$ satisfies (i), (ii) and (iii).

(4.9) Lemma. Let G, G_o, B be as in (4.8). Then the mapping $\pi : (x,y) \longmapsto xy$ is a 1-1 map of $G_o \times B$ onto G.

(4.10) Lemma. Let X be a locally compact space such that every open subset of X is σ-compact. Let ν be a positive and countably additive Borel measure on X such that $\nu(F) < \infty$ for compact $F \subseteq X$. Then ν is a regular

Borel measure on X in the sense of Hewitt and Ross [1,(11.34)].

The proof is left to the reader.

(4.11) Lemma. Notation is as in (4.8). If $D \subseteq G_o$ and $E \subseteq B$ are Borel sets in G_o and G respectively, then DE is a Borel set.

Proof. We show first that DE is a Suslin set in G. Indeed, the set $D \times E$ is a Borel rectangle in $G_o \times G$, so is a Borel and hence a Suslin set. The set DE is the image of $D \times E$ under the continuous map $(x,y) \longmapsto xy$ of $G_o \times G$ into G, and is thus a Suslin set in G (Bourbaki [2,Ch.IX,§6,Prop.11,Cor.]).

Now, in view of lemma (4.9) it is easy to see that the four sets DE, $(G_o \cap D')E$, $D(B \cap E')$, and $(G_o \cap D')(B \cap E')$ form a partition of G into disjoint sets. Since D and $G_o \cap D'$ are Borel sets in G_o, while E and $B \cap E'$ are Borel subsets of G, it follows from the preceding paragraph that all four sets DE, $(G_o \cap D')E$, $D(B \cap E')$, and $(G_o \cap D')(B \cap E')$ are Suslin sets. Therefore DE is a Borel set (Bourbaki, loc. cit. Thm. 2, Cor.).

(4.12) Lemma. Notation is as in (4.8). For each compact set K that is a neighborhood of e in G, there is a real number $\alpha > 0$ such that

$$\lambda(P((G_oK) \cap B)) = \alpha\lambda_o(P) \quad \text{for all Borel sets } P \subseteq G_o$$

where λ, λ_o denote left Haar measure on G, G_o respectively.

Proof. Let K be given and for each Borel set $P \subseteq G_o$ set

$$\nu(P) = \lambda(P((G_oK) \cap B)).$$

We shall show first that ν is a countably additive positive Borel measure on G_o. If $P \subseteq G_o$ is Borel, then since G_oK is a closed and B is Borel, lemma (4.11) shows that $P((G_oK) \cap B)$ is a Borel set, so that ν is certainly well defined; ν is clearly positive.

Now let $\{P_n\}_{n=1}^{\infty}$ be a countable disjoint family of Borel subsets of G_o. Then for $n > 0$, $P_n((G_oK) \cap B)$ is a Borel subset of G. Since P_m, P_n are disjoint if $m \neq n$, (4.9) shows that $P_m((G_oK) \cap B)$ and $P_n((G_oK) \cap B)$ are likewise disjoint if $m \neq n$. Hence we have

$$\nu\left(\bigcup_{n=1}^{\infty} P_n\right) = \lambda\left(\left(\bigcup_{n=1}^{\infty} P_n\right)((G_oK) \cap B)\right)$$

$$= \lambda\left(\bigcup_{n=1}^{\infty} (P_n((G_oK) \cap B))\right)$$

$$= \sum_{n=1}^{\infty} \lambda(P_n((G_oK) \cap B))$$

$$= \sum_{n=1}^{\infty} \nu(P_n)$$

the second and third equalitites following respectively from (4.9) and countable additivity of λ. Thus ν is countably additive on Borel sets of G_o.

Let $K^* \subseteq G_o$ be compact. Then $K^*((G_oK) \cap B)$ is contained in the compact set $\overline{K^*((G_oK) \cap B)}$, and hence $\nu(K^*) < \infty$. In addition, $G_o((G_oK) \cap B) \supseteq K$, so that $\nu(G_o) \geq \lambda(K) > 0$, since K has nonvoid interior.

Using the definition of ν and translation invariance of λ, it is easily seen that ν is left translation invariant.

Since G_o is locally compact and has a countable base for its topology , every open subset of G_o inherits these properties, and so is locally compact and σ-compact. Thus the countably additive positive measure ν on G_o satisfies the conditions of lemma (4.10), and therefore ν is regular.

Thus by the uniqueness of left Haar measure on G_o, there is an $\alpha > 0$ such that

$$\nu(P) = \alpha\lambda_o(P)$$

for all Borel sets $P \subseteq G_o$. This proves the assertion.

(4.13) Lemma. Notation is as in (4.8).

(i) If $D \subseteq G_o$ and $\lambda_o(D) = 0$, then $\lambda(DE) = 0$ for each Borel set $E \subseteq B$.

(ii) If $D \subseteq G_o$ and D is λ_o-measurable, then DE is λ-measurable for each Borel set $E \subseteq B$.

Proof. (i) Let V_n be an open subset of G_o such that $\lambda_o(V_n) < \frac{1}{n}$ and $D \subseteq V_n$. Such a V_n exists by regularity of λ_o. Set $D_1 = \bigcap_{n=1}^{\infty} V_n$. Then $D \subseteq D_1$, D_1 is a Borel set, and we have

$$\begin{aligned} \lambda_o(D_1) &= \lambda_o\left(\bigcap_{n=1}^{\infty} V_n\right) \\ &= \inf\{\lambda_o(V_n) : n = 1,2,\ldots\} \\ &= 0. \end{aligned}$$

Let $\{K_n\}_{n=1}^{\infty}$ be an ascending sequence of compact subsets of G, each with nonvoid interior in G, and such that $G = \bigcup_{n=1}^{\infty} K_n$. Then we have

$$\begin{aligned} D_1B &= D_1\left[\bigcup_{n=1}^{\infty} ((G_oK_n) \cap B)\right] \\ &= \bigcup_{n=1}^{\infty} [D_1((G_oK_n) \cap B)] \end{aligned}$$

by virtue of (4.9), and hence we have by (4.12)

$$\begin{aligned} \lambda(DE) \leq \lambda(DB) &\leq \lambda(D_1B) \\ &= \sup\{\lambda(D_1((G_oK_n) \cap B)) : n = 1,2,\ldots\} \\ &= \sup\{\alpha_n \cdot \lambda_o(D_1) : n = 1,2,\ldots\} \\ &= 0 \end{aligned}$$

where α_n is chosen so that $\lambda(D_1((G_oK_n) \cap B)) = \alpha_n\lambda_o(D_1)$.

(ii) If $D \subseteq G_o$ is λ_o-measurable, then by regularity of λ_o there are disjoint sets D_1, D_2 such that D_1 is a Borel set, $\lambda_o(D_2) = 0$, and $D = D_1 \cup D_2$. If $E \subseteq B$ is Borel, then (4.9) shows that

$$\begin{aligned} DE &= (D_1 \cup D_2)E \\ &= (D_1E) \cup (D_2E) \end{aligned}$$

and (4.11) and (i) supra show respectively that D_1E is a Borel set and that $\lambda(D_2E) = 0$. Thus DE is λ-measurable.

(4.14) Lemma. Notation is as in (4.8). There is a closed neighborhood S of e in G such that $G_oS = S$ and $\overline{S \cap B}$ is compact.

Proof. Let U be an open neighborhood of e in G such that $\overline{U}$ is compact. Set $S = G_o\overline{U}$. Then S is a closed neighborhood of G_o such that $G_oS = G_o(G_o\overline{U}) = G_o\overline{U} = S$. Also we have

$$\overline{S \cap B} = \overline{(G_o\overline{U}) \cap B}$$

which is compact by (4.8.ii).

(4.15) Lemma. Notation is as in (4.8) and (4.9). If $u \in G$, denote the preimage of u under the mapping π by $(\gamma(u),\beta(u))$, so that γ, β are functions from G onto G_o and B, respectively. We have the following propositions.

(i) If $u \in G$ then $u = xy$ with $x \in G_o$, $y \in B$ if and only if $x = \gamma(u)$, $y = \beta(u)$,

(ii) $\beta(u)$ is the unique element of $(G_ou) \cap B$, for each $u \in G$,

(iii) if $P \subseteq G_o$, then $\gamma^{-1}(P) = PB$,

(iv) $\gamma(ax) = a\gamma(x)$ for $a \in G_o$, $x \in G$.

(4.16) Lemma. Notation is as in (4.8) through (4.15). For each function f on G_o with complex values set

$$^S f(x) = f(\gamma(x)) \cdot \xi_S(x)$$

for all $x \in G$. Then $^S f$ is a complex-valued function on G with the following properties.

(i) If f is Haar measurable on G_o, then $^S f$ is Haar measurable on G.

(ii) If f has compact support, so does $^S f$.

(iii) There is a number $M > 0$ such that for $1 \leq p < \infty$ the mapping $f \longmapsto M^{1/p} \cdot {}^S f$ is a linear isometry of $L^p(G_o)$ into $L^p(G)$.

(iv) The mapping $f \longmapsto {}^S f$ is a linear isometry of $L^\infty(G_o)$ into $L^\infty(G)$.

(v) Let $1 < p < \infty$ and let $p' = \frac{p}{p-1}$. If $f \in L^p(G_o)$ and $g \in L^{p'}(G_o)$ then for all $x \in G_o$ we have

$$(*) \qquad M\left[{}^S f * \overline{({}^S g)^\star}\right](x) = f * \overline{g}^\star(x),$$

the convolutions on the left and right being over G and G_o respectively.

(vi) Let $f \in L^1(G_o)$ and let $g \in L^\infty(G_o)$, or let $f \in L^\infty(G_o)$ and let $g \in L^1(G_o)$. Then for all $x_o \in G$, $(*)$ holds.

Proof of (i). Suppose first that f is real-valued. Clearly $^S f$ is then real-valued, and we have for all real α

$$\{x \in G : f \circ \gamma(x) > \alpha\} = \{x \in G : \gamma(x) \in \{y \in G_o : f(y) > \alpha\}\}$$
$$= \left(\{y \in G_o : f(y) > \alpha\}\right)B$$

by (4.15.iii). Since f is by assumption measurable, the set $\{y \in G_o : f(y) > \alpha\}$ is λ_o-measurable. Then (4.13.ii) shows that

$(\{y \in G_o : f(y) > \alpha\})B$ is λ-measurable. Thus f is measurable. Since S is a closed set in G, ξ_S is a measurable function. Thus ${}^S f = (f \circ \gamma)\xi_S$ is measurable.

Now let f be an arbitrary measurable complex-valued function. A standard argument on $\mathcal{R}e\, f$ and $\mathcal{I}m\, f$ shows that again ${}^S f$ is measurable.

Proof of (ii). If ${}^S f(x) \neq 0$, then $f \circ \gamma(x) \neq 0$ and $\xi_S(x) \neq 0$. Thus we have by (4.15.iii)

$$\operatorname{supp}({}^S f) \subseteq \overline{\gamma^{-1}(\operatorname{supp} f) \cap S} = \overline{[(\operatorname{supp} f) \cdot B] \cap S}\ .$$

Now, if $y \in [(\operatorname{supp} f) \cdot B] \cap S$ then $\gamma(y) \in \operatorname{supp} f$ and $\beta(y) = \gamma(y)^{-1} y \in G_o S = S$. Thus $y \in (\operatorname{supp} f) \cdot (B \cap S)$. This shows that

$$\operatorname{supp}({}^S f) \subseteq \overline{[(\operatorname{supp} f) \cdot B] \cap S} \subseteq (\operatorname{supp} f) \cdot \overline{(B \cap S)},$$

the latter set being compact since supp f and $\overline{B \cap S}$ are compact.

Proof of (iii). It will be convenient to prove the following fact first. Let f be a real-valued measurable function on G_o, and let $\delta > 0$: if $\lambda_o(\{x_o \in G_o : f(x) > \delta\}) > 0$ then $\lambda(\{y \in G : {}^S f(y) > \delta\}) > 0$. Indeed, let K_o be a compact neighborhood of e in G contained in S, and set $P = \{x_o \in G_o : f(x_o) > \delta\}$. Then since $\lambda_o(P) > 0$, (4.12) shows that $\lambda(P((G_o K_o) \cap B)) > 0$. Because $K_o \cap S$ and $P \subseteq S$, we have

$$P((G_o K_o) \cap B) \subseteq P(S \cap B) \subseteq (PB) \cap S$$

so that $\lambda((PB) \cap S) > 0$ (we are not concerned with the measurability of $(PB) \cap S$). Now note that $y \in (PB) \cap S$ implies that $\gamma(y) \in P$ and $y \in S$. This shows that $(PB) \cap S \subseteq \{y \in G : {}^S f(y) < \delta\}$, and hence that the latter set has positive λ-measure, proving the claim.

Now let $f \in C_{oo}^{+}(G_o)$. Then f is λ_o-measurable, so that ^{S}f is bounded, λ-measurable by (i) of the present theorem, and has compact support by (ii). Thus we may define

$$I(f) = \int_G {}^{S}f\, d\lambda,$$

obtaining a complex-valued, positive homogeneous and additive function on $C_{oo}^{+}(G)$. If $f \neq 0$ then for some $\delta > 0$ the set $\{x_o \in G_o : f(x_o) > \delta\}$ is nonvoid and open in G_o; the previous paragraph then shows that $\lambda(\{y \in G : {}^{S}f(y) > \delta\}) > 0$, implying that $I(f) > 0$. Finally, if $a \in G_o$ and $x \in G$, we have $\xi_S(x) = \xi_S(ax)$, which shows in conjunction with (4.15.iv) and an otherwise routine calculation that

$$I({}_af) = \int_G {}^{S}({}_af)d\lambda = I(f).$$

Thus I is left invariant on $C_{oo}^{+}(G_o)$. It follows by the uniqueness theorem for the left Haar integral (Hewitt and Ross [1,(15.5)] that there is a constant $M > 0$ such that for $f \in C_{oo}^{+}(G)$ we have

$$\int_{G_o} f(x)d\lambda_o(x) = M \cdot I(f),$$

in other words

$$\int_{G_o} f(x)d\lambda_o(x) = M\int_G {}^{S}f(x)d\lambda(x) \tag{1}$$

holds for $f \in C_{oo}^{+}(G_o)$. Standard arguments, which we omit, may be used to deduce that formula (1) actually holds for all λ_o-measurable functions f on G_o. The remainder of (iii) has a routine proof.

Proof of (iv). Since G, G_o are σ-compact, the distinction between locally null and null sets disappears. Let $f \in L^{\infty}(G_o)$. Then the set $\{x \in G_o : |f(x)| > \delta\}$ is nonnull for each $\delta < ||f||_{\infty}$. Thus by the fact proved at the beginning of the proof of (iii) we know that $\{y \in G : {}^S|f|(x) > \delta\}$ is also nonnull. Since ${}^S|f| = |{}^Sf|$ this implies that $\delta < \|{}^Sf\|_\infty$. It follows that we have $\|f\|_\infty \leq \|{}^Sf\|_\infty$. Conversely, let $\delta < \|{}^Sf\|_\infty$. Then there is a set $Q \subseteq G$ such that $\lambda(Q) > 0$ and $|{}^Sf(y)| > \delta$ for $y \in Q$. Then $Q \subseteq S$ and we have $|f(y)| > \delta$ for $y \in \gamma(Q)$. Then since $Q \subseteq \gamma(Q)\cdot B$ (see (4.15.i)) we have $0 < \lambda(Q) < \lambda(\gamma(Q)\cdot B)$ and therefore $\lambda(\gamma(Q)) > 0$ by (4.13.i). Therefore $\{x \in G_o : f(x) > \delta\}$ contains $\gamma(Q)$ which has positive measure, implying the inequality $\delta < \|f\|_\infty$. This proves that $\|{}^Sf\|_\infty \leq \|f\|_\infty$, which in turn completes the proof of (iv).

<u>Proof of</u> (v) <u>and</u> (vi). Under these assumptions, it is an immediate consequence of (iii) and (iv) <u>supra</u> that the pairs $f, \overline{g^*}$ and ${}^Sf, \overline{({}^Sg)^*}$ satisfy the conditions of theorem (20.16) of Hewitt and Ross [1], so that the convolutions in the formula (*) each make sense and are defined for all $x \in G_o$, when interpreted according to that theorem. Noting (4.15.iv) and that $\xi_S(xy)\xi_S(y) = |\xi_S(y)|^2 = \xi_S(y)$ for $x \in G_o$, $y \in G$, proof of formula (*) is immediate.

(4.17) <u>Lemma</u>. Let G be a locally compact group and let N be a compact normal subgroup of G. Let λ_N denote normalized Haar measure on N, regarded as an element of $\underline{\underline{M}}(G)$ in the usual way. Then the following propositions hold.

(i) If $f \in L^p(G)$ $(1 \leq p < \infty)$ then $\lambda_N * f$ and $f * \lambda_N$ are in $L^p(G)$, and each of the latter two functions is equal almost everywhere to a function constant on cosets of N.

(ii) If either $f \in L^p(G)$, $1 \leq p < \infty$, or $f \in C_o(G)$, and if f is constant on cosets of N, then f, $\lambda_n * f$, and $f * \lambda_N$ are equal almost everywhere.

(iii) If $1 < p < \infty$, $p' = \frac{p}{p-1}$, $g \in L^p(G)$, and $h \in L^{p'}(G)$, then we have

$$(1)\quad \lambda_N * (g * h^*) = (\lambda_N * g) * h^* ,$$

$$(2)\quad (g * h^*) * \lambda_N = g * (h^* * \lambda_N),$$

and

$$(3)\quad h^* * \lambda_N = (\lambda_N * h)^* .$$

The next lemma follows readily from Hewitt and Ross [2,(28.54)].

(4.18) <u>Lemma</u>. Let G be a locally compact group, let N be a compact normal subgroup of G, let σ denote the canonical homomorphism of G onto G/N, and let $1 < p < \infty$. Then in order that a Borel measurable function f on G be almost everywhere equal to $\tilde{f} \circ \sigma$ for some Borel measurable function $\tilde{f} \in L^p(G/N)$, it is necessary and sufficient that f be in $L^p(G)$ and be equal almost everywhere to a function constant on each coset of N. Haar measure λ_1 on G/N may be chosen so that if f and $\tilde{f}$ are related in this way, then $||\tilde{f}||_p = ||f||_p$.

(4.19) <u>Theorem</u>. Let G be a locally compact group, let N be a compact normal subgroup of G, and let σ denote the canonical homomorphism $x \longmapsto xN$ of G onto G/N. If $1 < p < \infty$, then in order that a function f on G have the form $\tilde{f} \circ \sigma$ for some $\tilde{f} \in A_p(G/N)$, it is necessary and sufficient that $f \in A_p(G)$ and f be constant on cosets of N.

<u>Proof</u>. Let $\tilde{f} \in A_p(G/N)$. Then certainly $f = \tilde{f} \circ \sigma$ is constant on cosets of N. We may write

$$\tilde{f} = \sum_{k=1}^{\infty} g_k * h_k^*$$

for suitable $g_k \in L^p(G/N)$ and $h_k \in L^{p'}(G/N)$, g_k and h_k being chosen to be

Borel measurable functions. Then we have by (4.18) that for suitable choice λ_1 of Haar measure on G/N we have

$$||g_k \circ \sigma||_p = ||g_k||_p, \quad k = 1,2,\ldots,$$

and

$$||h_k \circ \sigma||_{p'} = ||g_k||_{p'}, \quad k = 1,2,\ldots .$$

Thus, for $k = 1,2,\ldots,$ the function $(g_k \circ \sigma) * (h_k \circ \sigma)^{\star}$ is defined as in Hewitt and Ross [1,(20.16)] and we have for all $x \in G$

$$(g_k \circ \sigma) * (h_k \circ \sigma)^{\star}(x) = (g_k * h_k^{\star}) \circ \sigma(x)$$

where the calculation uses Hewitt and Ross [2,(28.54)] in conjunction with membership of $[_{\sigma(x)}g_k][h_k]$ in $L^1(G/N)$. Now we have

$$\begin{aligned}\sum_{k=1}^{\infty} ||g_k \circ \sigma||_p \, ||h_k \circ \sigma||_{p'} &= \sum_{k=1}^{\infty} ||g_k||_p \, ||h_k||_{p'} \\ &= \sum_{k=1}^{\infty} ||g_k||_p \, ||h_k||_{p'} \\ &= \infty,\end{aligned}$$

showing that the sum $\sum_{k=1}^{\infty} (g_k \circ \sigma) * (h_k \circ \sigma)^{\star}$ defines a function in $A_p(G)$, and the previous calculation shows that

$$\begin{aligned}\sum_{k=1}^{\infty} (g_k \circ \sigma) * (h_k \circ \sigma)^{\star} &= \left(\sum_{k=1}^{\infty} g_k * h_k^{\star}\right) \circ \sigma \\ &= \tilde{f} \circ \sigma \\ &= f.\end{aligned}$$

This proves that indeed $f \in A_p(G)$. The necessity of the stated condition is thus proved.

Now, let $f \in A_p(G)$ be constant on cosets of N. Then we have by (4.17.ii) $\lambda_N * f = f * \lambda_N = f$. We also have $f = \sum_{k=1}^{\infty} g_k * h_k^{\star}$ for suitable Borel measurable functions $g_k \in L^p(G)$, $h_k \in L^{p'}(G)$. Using (4.17.iii) and taking into account the uniform absolute convergence of the sum representing f, we have

$$\begin{aligned} f &= (\lambda_N * f) * \lambda_N \\ &= \left[\lambda_N * \left(\sum_{k=1}^{\infty} g_k * h_k^{\star}\right)\right] * \lambda_N \\ &= \sum_{k=1}^{\infty} (\lambda_N * g_k) * (h_k^{\star} * \lambda_N) \\ &= \sum_{k=1}^{\infty} (\lambda_N * g_k) * (\lambda_N * h_k)^{\star} . \end{aligned}$$

Since $\lambda_N * g_k \in L^p(g)$ and is constant on cosets of N, we may invoke (4.18) to write $\lambda_N * g_k = \tilde{g}_k \circ \sigma$ almost everywhere, for some $\tilde{g}_k \in L^p(G/N)$. Similarly we may write $\lambda_N * h_k = \tilde{h}_k \circ \sigma$ almost everywhere, with $\tilde{h}_k \in L^{p'}(G/N)$. Then we have by (4.18) that

$$\begin{aligned} \|\tilde{g}_k\|_p &= \|\lambda_N * g_k\| \\ &\leq \|\lambda_N\| \ \|g_k\|_p \\ &= \|g_k\|_p \end{aligned}$$

and similarly we have

$$\|\tilde{h}_k\|_{p'} \leq \|h_k\|_{p'} ,$$

and so we see that

$$\sum_{k=1}^{\infty} \|\tilde{g}_k\|_p \, \|\tilde{h}_k\|_{p'} \le \sum_{k=1}^{\infty} \|g_k\|_p \, \|h_k\|_{p'} < \infty .$$

Thus the sum $\sum_{k=1}^{\infty} \tilde{g}_k * (\tilde{h}_k)^{*}$ represents a function $\tilde{f}$ in $A_p(G/N)$. For all $x \in G$ we have, by Hewitt and Ross [2,(28.54)] and the above calculations,

$$\begin{aligned} \tilde{f} \circ \sigma(x) &= \sum_{k=1}^{\infty} (\tilde{g}_k \circ \sigma) * (\tilde{h}_k \circ \sigma)^{*}(x) \\ &= \sum_{k=1}^{\infty} (\lambda_N * g_k) * (\lambda_N * h_k)^{*}(x) \\ &= f(x) . \end{aligned}$$

This proves the sufficiency of the stated condition.

(4.20) Lemma. Let G be a locally compact group, let G_1 be an open subgroup of G, and let $1 < p < \infty$. If $f' \in A_p(G_1)$, then there is an extension $f \in A_p(G)$ of f' that vanishes outside G_1. If $p = 2$ and f' is also positive-definite, we may choose f to be positive-definite.

Proof. We first make a remark concerning Haar measure on G_1. Let λ, λ' denote Haar measures for G, G_1 respectively. Since G_1 is open in G, we have $\lambda(G_1) > 0$, and therefore λ, when restricted to act on subsets of G_1, satisfies the conditions of the uniqueness theorem for Haar measure on G_1. Hence there exists a number $c > 0$ such that $d\lambda = c d\lambda'$ on G_1.

Now we proceed to the proof of the lemma. Let $f' \in A_p(G_1)$, so that $f' = \sum_{k=1}^{\infty} g'_k * (h'_k)^{*}$ with $g'_k \in L^p(G_1)$ and $h'_k \in L^{p'}(G_1)$, $k = 1,2,\ldots$.
Extend g'_k and h'_k by setting

$$g_k(x) = g_k'(x), \quad x \in G_1$$

$$h_k(x) = h_k'(x), \quad x \in G_1$$

$$g_k(x) = h_k(x) = 0, \quad x \notin G_1.$$

Thus with c as in the first paragraph of this proof we have

$$\|g_k\|_p = c^{1/p}\|g_k'\|_p ,$$

and

$$\|h_k\|_{p'} = c^{1/p'}\|h_k'\|_{p'} .$$

Therefore we have

$$\sum_{k=1}^{\infty} \|g_k\|_p \, \|h_k\|_{p'} = c \sum_{k=1}^{\infty} \|g_k'\|_p \, \|h_k'\|_{p'} < \infty$$

so that the sum $\sum_{k=1}^{\infty} \frac{1}{c} g_k * h_k^{\star}$ is a function f in $A_p(G)$. Noting that for $x \in G_1$ we have $xy \in G_1$ if and only if $y \in G_1$, we may also state that for all $x \in G_1$ we have

$$g_k * h_k^{\star} = c(g_k' * (h_k')^{\star})(x),$$

and thus for $x \in G_1$ we have

$$f(x) = \sum_{k=1}^{\infty} \frac{1}{c} g_k * h_k^{\star}(x)$$

$$= \sum_{k=1}^{\infty} g_k' * (h_k') \ (x)$$

$$= f'(x).$$

This shows that f extends f'.

If $x \notin G_1$ then we have

$$f(x) = \frac{1}{c} \sum_{k=1}^{\infty} g_k * h_k^{\star}(x)$$

$$= \frac{1}{c} \sum_{k=1}^{\infty} \int_G g_k(xy)h_k(y)d\lambda(y)$$

$$= 0,$$

since if $x \notin G_1$ then for all $y \in G$, either $y \notin G_1$ or $xy \notin G_1$.

Finally, if $p = 2$ and f' is positive-definite, then $f' = g_1' * \overline{(g_1')^{\star}}$ with $g_1' \in L^2(G_1)$, by (4.2) and (4.4). Defining g_1 in terms of g_1' as above and applying the calculations above we see that $\frac{1}{c} g_1 * \overline{g_1^{\star}}$ is an extension of f' and is positive-definite by (4.4).

(4.21) Theorem. Let G be a locally compact group, and let G_o be a closed subgroup of G. If $1 < p < \infty$ and if $f \in A_p(G_o)$, then f has an extension in $A_p(G)$. If $p = 2$ and f is also positive-definite, so that $f = g * \overline{g^{\star}}$ with $g \in L^2(G_o)$ by (4.4), then f has an extension in $A(G)$ of the form $g' * \overline{(g')^{\star}}$ with $g' \in L^2(G)$.

Proof. (a) Consider first the case when G has a countable base for its topology. Since $f \in A_p(G_o)$, we may write

$$f = \sum_{k=1}^{\infty} g_k * \overline{h_k^{\star}}, \quad \sum_{k=1}^{\infty} \|g_k\|_p \, \|h_k\|_{p'} < \infty,$$

where g_k, h_k are suitable functions in $L^p(G_o)$, $L^{p'}(G_o)$ respectively. Then we have for each k, by (4.16.iii),

$$\|M^S g_k\|_p \, \|{}^S h_k\|_{p'} = \|M^{1/p}\, {}^S g_k\|_p \, \|M^{1/p'}\, {}^S h_k\|_{p'} = \|g_k\|_p \, \|h_k\|_{p'},$$

and therefore

$$\sum_{k=1}^{\infty} \| M^S g_k \|_p \; \| {}^S h_k \|_{p'} < \infty ,$$

so that the sum $\sum_{k=1}^{\infty} M^S g_k * \overline{({}^S h_k)^{\star}}$ is function f' in $A_p(G)$. For $x \in G_o$ we have by (4.16.v)

$$f'(x) = \sum_{k=1}^{\infty} M\left[{}^S g_k * \overline{({}^S h_k)^{\star}}\right](x)$$

$$= \sum_{k=1}^{\infty} g_k * \overline{h_k^{\star}}(x)$$

$$= f(x),$$

showing that f' extends f. If $p = 2$ and $f = g * \overline{g^{\star}}$ with $g \in L^2(G)$, the above calculations show that $f' = (M^{1/2}\, {}^S g) * \overline{(M^{1/2}\, {}^s g)^{\star}}$ extends f.

(b) Now suppose that G is σ-compact and locally compact. Since G is a normal topological space there is a continuous function g on G extending f. Since g is then a fortiori a Baire function on G, Kakutani and Kodaira [1,Satz 6] assure us that there is a compact normal subgroup N of G such that g is constant on cosets of N, and G/N has a countable base for its topology.

Suppose that $x, y \in G_o$ and $xy^{-1} \in G_o \cap N$. Then $xy^{-1} \in N$, and hence $f(x) = g(x) = g(y) = f(y)$. That is to say, f is constant on cosets of $G_o \cap N$ in G_o.

Now, since G_o is closed and N is compact, G_oN is closed and hence is locally compact and σ-compact. Theorems (5.32) and (5.33) of Hewitt and Ross [1] show that there is a topological isomorphism τ of G_oN/N onto $G_o/(G_o \cap N)$ defined by setting $\tau(xN) = x(G_o \cap N)$ for $x \in G_o$. By (4.19) there is a function $\tilde{\tau} \in A_p(G_o/G_o \cap N))$ such that $f = \tilde{f} \circ \sigma$, σ being the canonicial quotient homomorphism from G_o onto $G_o/(G_o \cap N)$. Because

$(G_oN)/N$ is locally compact, it is a closed subgroup of G/N (Hewitt and Ross [1,(5.11)]). Applying case (a) to the function $f' = \tilde{f} \circ \tau \in A_p((G_oN)/N)$, which is legitimate since G/N has a countable base for its topology, we deduce the existence of a function $f'' \in A_p(G/N)$ extending f'. Applying (4.19) again, we see that $f''' = f'' \circ \sigma' \in A_p(G)$, σ' denoting the natural quotient homomorphism from G onto G/N. For $x \in G_o$ we then have

$$\begin{aligned} f'''(x) &= f''(xN) \\ &= f'(xN) \\ &= (\tilde{f} \circ \tau)(xN) \\ &= \tilde{f}(x(G_o \cap N)) \\ &= (\tilde{f} \circ \sigma)(x) \\ &= f(x) \end{aligned}$$

showing that f''' extends f. If $p = 2$ and f is positive-definite, then part (a) ensures that $\tilde{f}$, f', f'' and f''' in the above proof may be assumed to be positive-definite, and thus that f''' has the required form by (4.4).

(c) Now let G be an arbitrary locally compact group, and let $f = \sum_{k=1}^{\infty} g_k * \overline{h_k^{\star}}$, with $g_k \in L^p(G_o)$, $h_k \in L^{p'}(G_o)$ and $\sum_{k=1}^{\infty} \|g_k\|_p \|h_k\|_{p'} < \infty$.

We prove first the existence of a σ-compact open subgroup G_1 of G, such that for $k \geq 1$ and almost all $x \in G_1'$ we have $g_k(x) = 0$ and $h_k(x) = 0$. Indeed, $C_{oo}(G_o)$ is dense in $L^p(G_o)$ and in $L^{p'}(G_o)$. Therefore, for $k, n = 1,2,\ldots$ there is a function $\varphi_{kn} \in C_{oo}(G)$ such that on G_o

$$\|g_k - \varphi_{kn}\|_p < \frac{1}{n}$$

and a function $\psi_{kn} \in C_{oo}(G_o)$ such that on G_o

$$\|h_k - \psi_{kn}\|_{p'} < \frac{1}{n} .$$

Let C_1 be any relatively compact neighborhood of e in G. Since for each $m > 1$ the set $\bigcup\{\text{supp } \varphi_{kn} : k,n \leq m\} \cup \bigcup\{\text{supp } \psi_{kn} : k,n \leq m\}$ is compact, there is a relatively compact open symmetric neighborhood C_m of e in G containing it and C_{m-1}. Let $L_m = \bigcup_{j=1}^{\infty} C_m^j$, so that L_m is a σ-compact open subgroup of G. Clearly $L_m \subseteq L_{m+1}$ for $m \geq 1$, since $C_m \subseteq C_{m+1}$ for $m \geq 1$, and so $G_1 = \bigcup_{m=1}^{\infty} L_m$ is an open σ-compact subgroup of G. Each φ_{kn} and each ψ_{kn} vanish on $G_o \cap G_1'$, since supp φ_{kn} and supp $\psi_{kn} \subseteq G_1$ for all k,n. Therefore we have for $k,n \geq 1$, λ_o denoting Haar measure on G_o,

$$\begin{aligned}
\left\{\int_{G_o \cap G_1'} |g_k|^p d\lambda_o\right\}^{1/p} &\leq \left\{\int_{G_o \cap G_1'} |g_k - \varphi_{kn}|^p d\lambda_o\right\}^{1/p} \\
&\quad + \left\{\int_{G_o \cap G_1'} |\varphi_{kn}|^p d\lambda_o\right\}^{1/p} \\
&\leq \left\{\int_{G_o} |g_k - \varphi_{kn}|^p d\lambda_o\right\}^{1/p} \\
&= \| g_k - \varphi_{kn}\|_p \\
&< \frac{1}{n}
\end{aligned}$$

and since this holds for arbitrary $n \geq 1$ we have $g_k(x) = 0$ for almost all $x \in G_o \cap G_1'$. An analogous calculation produces the same conclusion for h_k.

Set $f' = f|G_o \cap G_1$, $g_k' = g_k|G_o \cap G_1$, $h_k' = h_k|G_o \cap G_1$. Then we have on $G_o \cap G_1$

$$\begin{aligned}
\| g_k'\|_p &= \left\{\int_{G_o \cap G_1} |g_k'|^p d\lambda'\right\}^{1/p} \\
&= c^{-\frac{1}{p}} \left\{\int_{G_o \cap G_1} |g_k'|^p d\lambda_o\right\}^{1/p}
\end{aligned}$$

$$= c^{-\frac{1}{p}} \left\{ \int_{G_o \cap G_1} |g_k|^p d\lambda_o \right\}^{1/p}$$

$$= c^{-\frac{1}{p}} \left\{ \int_{G_o} |g_k|^p d\lambda_o \right\}^{1/p}$$

$$= c^{-\frac{1}{p}} \|g_k\|_p$$

where λ_o, λ' denote Haar measure on G_o and its open subgroup $G_o \cap G_1$, respectively, and where c is a constant such that $d\lambda_o = c d\lambda'$ (see the remark beginning the proof of (4.20)). Similarly we have

$\|h'_k\|_p = c^{-\frac{1}{p}} \|h_k\|_{p'}$. Thus $\sum_{k=1}^{\infty} \|g'_k\|_p \|h'_k\|_{p'} = \sum_{k=1}^{\infty} c^{-1} \|g_k\|_p \|h_k\|_p < \infty$,

so that the sum

$$\sum_{k=1}^{\infty} g'_k * \overline{(h'_k)^{\star}}$$

is a function in $A_p(G_o \cap G_1)$. However, for $x \in G_o \cap G_1$ we have

$$c \sum_{k=1}^{\infty} g'_k * \overline{(h'_k)^{\star}}(x) = c \sum_{k=1}^{\infty} \left\{ \int_{G_o \cap G_1} g'_k(xy)\overline{h'_k}(y) d\lambda'(y) \right\}$$

$$= \sum_{k=1}^{\infty} \left\{ \int_{G_o \cap G_1} g_k(xy)\overline{h_k}(y) d\lambda(y) \right\}$$

$$= \sum_{k=1}^{\infty} \left\{ \int_{G_o} g_k(xy)\overline{h_k}(y) d\lambda_o(y) \right\}$$

$$= \sum_{k=1}^{\infty} g_k * \overline{h_k^{\star}}(x)$$

$$= f'(x)$$

and therefore f' is a member of $A_p(G_o \cap G_1)$. Since $G_o \cap G_1$ is closed in G_1 and G_1 is locally compact and σ-compact, we may apply part (b) to deduce existence of an extension $f_1 \in A_p(G_1)$ for f'. Now we may further apply (4.20) to f_1 on the open subgroup G_1 of G, to obtain an extension f'' for f_1, with $f'' \in A_p(G)$ and $f''(x) = 0$ for $x \notin G_1$. The function f'' clearly extends f.

If $p = 2$, f is positive-definite and $f = g * \overline{g^\star}$ with $g \in L^2(G_o)$, then the last sentences both of (b) above and of (4.20) ensure that f_1 and f'' may be chosen to be likewise positive-definite.

The foregoing result gives an alternative proof of Theorem (2.7). Actually, we may now state an even stronger result.

(4.22) Corollary. Let G be a locally compact group and let G_o be a compact subgroup of G. Then every function $f \in \underline{\underline{P}}(G_o)$ has an extension to all of G of the form $g * \overline{g^\star}$, with $g \in L^2(G)$; a *fortiori* f has an extension that is positive-definite and vanishes at infinity.

Proof. Since $A(G_o) = \underline{\underline{K}}(G_o)$, as Hewitt and Ross [2,(34.16)] show, this corollary is a consequence of theorems (4.4), (4.21) and (4.3) *supra*.

* * *

(4.23) In view of the remark made in (3.11), the following consequence of (4.21) is also of some interest. Recall that a group G_o is said to be *amenable* if it has a nontrivial invariant mean for continuous bounded functions. There are in the literature many conditions equivalent to this one. The one we need for the present result is this (see Reiter [1]):

> (R): Every positive definite continuous function on G_o is the limit, uniformly on compact sets, of functions of the form $f * \overline{f^\star}$ with $f \in L^2(G)$.

Theorem. Let G be a locally compact group with an amenable closed subgroup G_o. Then in order that $\underline{\underline{P}}(G_o, G)$ be closed in $\underline{\underline{K}}(G_o)$ for the topology of compact convergence on G_o, it is necessary and sufficient that $\underline{\underline{P}}(G_o,G) = \underline{\underline{P}}(G_o)$.

Proof. Since $\underline{\underline{P}}(G_o)$ is closed in $\underline{\underline{K}}(G_o)$ for the topology of compact convergence, the sufficiency is trivial.

To prove the necessity, we note that condition (R) above implies that the closure of $A(G_o) \cap \underline{\underline{P}}(G_o)$ for the topology of compact convergence is precisely the set $\underline{\underline{P}}(G_o)$, by (4.4). Therefore, by (4.21), the closure of $\underline{\underline{P}}(G_o,G)$ for this topology is likewise equal to $\underline{\underline{P}}(G_o)$.

§5. EXTENSION FROM ABELIAN NORMAL SUBGROUPS

The results of this section were largely inspired by a result of Douady presented in Hewitt and Ross [2,(34.48.e)]. His theorem, therefore, appropriately begins our discussion.

Throughout §5, G will denote a locally compact group, A will denote a closed abelian normal subgroup of G, and "character" will mean "continuous character", unless the contrary is explicitly stated.

(5.1) Definition. For each character χ of A, by the stationary subgroup of A we mean the set

$$F(\chi) = \{y \in G : \chi(yxy^{-1}) = \chi(x) \text{ for all } x \in A\}.$$

The following lemma justifies this terminology. The proof is obvious, and is omitted.

(5.2) Lemma. The set $F(\chi)$ is a closed subgroup of G containing A.

(5.3) Douady's Theorem. For each character χ on A, if $\chi \in \underline{\underline{P}}(A,G)$ then $F(\chi)$ is open in G.

Proof. We need merely note that this is the contrapositive of the proposition *loc. cit*.

(5.4) *Corollary*. For every character χ of A, χ extends to G as an element of $\underline{\underline{P}}(G)$ if and only if $F(\chi)$ is open in G and χ extends to $F(\chi)$ as an element of $\underline{\underline{P}}(F(\chi))$.

Proof. By virtue of the fact that an open subgroup is always extending, this corollary is immediate.

(5.5) *Corollary*. Let G be a locally compact group. If A is an extending normal closed abelian subgroup of G, then A and the connected component C of G commute elementwise.

Proof. Suppose that $y \in C$ and $x \in A$ do not commute. Let χ be a continuous character of A such that $\chi(yxy^{-1}) \neq \chi(x)$. Then y is not in $F(\chi)$. Since $F(\chi)$ is open by (5.3), we have $C \subseteq F(\chi)$. Thus we have arrived at a contradiction.

(5.6) *Corollary*. Let G be a connected locally compact group. An extending closed normal abelian subgroup A of G is necessarily contained in the centre of G.

Proof. This is the instance of (5.5) when $C = G$.

(5.7) *Corollary*. Let G be a locally compact group that contains a closed abelian normal subgroup A and a closed connected subgroup H and suppose that $G = A \circledS H$. Then A is extending if and only if $G = A \times H$.

Proof. Observe that the connected component of G contains H. Thus if A is extending, then by (5.5) A and H commute elementwise, so that we have $G = A \times H$.

Conversely, if $G = A \times H$ and if $\varphi_o \in \underline{\underline{P}}(A)$, then setting $\varphi(x,h) = \varphi_o(x)$ defines a function φ on G that plainly belongs to $\underline{\underline{P}}(G)$ and extends φ_o.

(5.8) Discussion. For our next result, we use the two concepts of nilpotence and solvability. We need first to make some remarks concerning solvable topological groups.

Let G be a solvable topological group, of (solvable) length n. That is to say, there is a chain $\{e\} = G_o \subseteq G_1 \subseteq \cdots \subseteq G_n = G$ of normal subgroups of G such that G_k/G_{k-1} is abelian for $1 \leq k \leq n$. For $1 \leq k \leq n$ the group $G_k/\overline{G_{k-1}}$ is algebraically isomorphic with the group $(G_k/G_{k-1})/(\overline{G_{k-1}}/G_{k-1})$, and thus is also abelian. The group $\overline{G_k}/\overline{G_{k-1}}$ contains the dense abelian subgroup $G_k/\overline{G_{k-1}}$, and so is in turn also abelian. Thus every solvable topological group G has a chain $\{e\} = \overline{G_o} \subseteq \overline{G_1} \subseteq \cdots \subseteq \overline{G_n} = G$ of closed normal subgroups such that for $1 \leq k \leq n$ the group $\overline{G_k}/\overline{G_{k-1}}$ is abelian.

(5.9) Theorem. Let G be a locally compact connected solvable group with the extension property. Then G is nilpotent.

Proof. We shall give a proof by induction on the solvable length n of G. The case $n = 1$ is trivial, since then G is abelian.

Now let G be a solvable group of length n, and suppose that every connected solvable group of length less than n is nilpotent. Let $\{e\} = G_o \subseteq G_1 \subseteq \cdots \subseteq G_n = G$ be a chain of closed normal subgroups of G such that G_k/G_{k-1} is abelian for $1 \leq k \leq n$. Then the chain (of length $n - 1$)

$$(G_1/G_1) \subseteq (G_2/G_1) \subseteq \cdots \subseteq (G_n/G_1)$$

is a chain of normal subgroups of the group G_n/G_1, with each quotient group $(G_k/G_1)/(G_{k-1}/G_1)$ abelian. Therefore, G/G_1 is solvable of length less than n. The group G/G_1 has the extension property by (3.5), and inherits connectedness from G. By the inductive hypothesis, then, G/G_1 is nilpotent.

Since the subgroup G_1 of G is extending, abelian, and normal, by (5.6) we see that G_1 is contained in the centre Z of G. The group G/Z, being topologically isomorphic with the quotient group $(G/G_1)/(Z/G_1)$ of G/G_1, is certainly nilpotent (loc.cit., VI.1.e). Therefore G is nilpotent (loc. cit. VI.1.h). This completes the induction proof.

Remark. Theorem (5.9) clearly implies that every connected solvable compact group is nilpotent. Hewitt and Ross [2,(29.44)] state a stronger result, viz that such a group is necessarily abelian. It is of interest, however, to note the completely different routes by which these two results are reached.

(5.10) Example. A locally compact connected nilpotent group need not have the extension property, as the following example shows.

Let G be the group of matrices of the form $\begin{bmatrix} 1 & 0 & 0 \\ u & 1 & 0 \\ w & v & 1 \end{bmatrix}$, with u, v, and w arbitrary real numbers, G being regarded as a subgroup of $GL(3,\mathbb{R})$. This group has been studied by Dixmier [1]. The group G is connected and nilpotent. Consider the set A of elements of G of the form $\begin{bmatrix} 1 & 0 & 0 \\ 0 & 1 & 0 \\ w & v & 1 \end{bmatrix}$, for any real y, y', z, and z' we have

$$\begin{bmatrix} 1 & 0 & 0 \\ 0 & 1 & 0 \\ w & v & 1 \end{bmatrix} \begin{bmatrix} 1 & 0 & 0 \\ 0 & 1 & 0 \\ w' & v' & 1 \end{bmatrix} = \begin{bmatrix} 1 & 0 & 0 \\ 0 & 1 & 0 \\ w+w' & v+v' & 1 \end{bmatrix}$$

showing at once that A is an abelian normal subgroup of G isomorphic with $\mathbb{R} \times \mathbb{R}$. The subgroup A is evidently closed in G. The elements

$$x_1 = \begin{bmatrix} 1 & 0 & 0 \\ 0 & 1 & 0 \\ 0 & 1 & 1 \end{bmatrix} \quad \text{and} \quad x_2 = \begin{bmatrix} 1 & 0 & 0 \\ 1 & 1 & 0 \\ 0 & 0 & 1 \end{bmatrix}$$

do not commute, as is readily checked. Thus A is not contained in the centre of G. By (5.6), then, A is not an extending subgroup of G.

(5.11) Discussion. Let G be a locally compact group, let X be a locally compact space, and let ι be a positive Borel measure on X. Suppose that for each $\alpha \in X$ we are given a cyclic representation $U^{(\alpha)}$ of G. Since any nonzero multiple of a cyclic vector is a cyclic vector, we may choose a cyclic vector $\zeta(\alpha)$ for $U^{(\alpha)}$ in such a way that the function $\alpha \longrightarrow \|\zeta(\alpha)\|$ is a continuous function belonging to $L^2(X,\iota)$. Godement [2], passim, has shown how the direct integral $U = \int_X \oplus U^{(\alpha)} d\iota$ may be defined. The representation space for U is the Hilbert space $L^2_{\Lambda,\iota}$ defined in loc. cit. for the measure ι and the particular "fundamental family" Λ consisting of all functions $\alpha \longmapsto U_x^{(\alpha)}\zeta(\alpha)$, as x runs through G. For each vector field $h \in L^2_{\Lambda,\iota}$ and for each $x \in G$, $U_x h$ is defined by the relation

$$(U_x h)(\alpha) = U_x^{(\alpha)}(h(\alpha)).$$

We shall call U the direct integral of the representations $U^{(\alpha)}$.

It is known (see Godement loc. cit.) that if G/Z(G) has a countable base for its topology, then every representation of G is a direct integral, in the sense defined above, of representations $U^{(\alpha)}$ which are nonzero and irreducible for ι-almost all α.

We can now justify (see (5.14) infra) the attention we are giving to extensions of mere characters of an abelian normal subgroup A of the group G. In view of (5.4), we may restrict our attention to characters χ of A which are invariant in G in the sense that $\chi(yxy^{-1}) = \chi(x)$ for all $x \in A$ and $y \in G$. That is, χ is invariant precisely when $F(\chi) = G$.

(5.12) Theorem. Let G be a locally compact group. Let A be a closed abelian normal subgroup of G, and let χ be an invariant character of A.

If $\chi \in \underline{\underline{\underline{P}}}(A,G)$, then there is a cyclic representation U of G that extends χ in such a way that we have

(i) $$U_x = \chi(x)I \quad \text{for} \quad x \in A,$$

where I denotes the identity operator on $H(U)$.

<u>Proof</u>. Let $\varphi \in \underline{\underline{\underline{P}}}(G)$ be an extension of χ, and let $U = V^{(\varphi)}$. Then we have $\chi(x) = \langle U_x\zeta, \zeta \rangle$ for $x \in A$, where ζ is a suitable cyclic vector for U. The argument used in Hewitt and Ross [2,(34.48.d)] shows that for $x \in A$ we have

$$U_x\zeta = \chi(x)\zeta.$$

Let x and y be elements of A and G respectively. Then we have

$$\begin{aligned} U_x(U_y\zeta) &= U_y(U_{y^{-1}xy})\zeta \\ &= U_y\chi(y^{-1}xy)\zeta \\ &= U_y\chi(x)\zeta \\ &= \chi(x)(U_y\zeta), \end{aligned}$$

and so the relation

$$U_x\xi = \chi(x)I\xi$$

holds for elements ξ of $H(U)$ which have the form $U_y\zeta$ for some $y \in G$. Thus this relation also holds for ξ in the closed linear span $H(U)$ of the set $\{U_y\zeta : y \in G\}$. This proves the claim.

(5.13) <u>Theorem</u>. Let G be a locally compact group with a countable base for its topology, and let A be a closed abelian normal subgroup of G. Let

χ be an invariant character in $\underline{\underline{P}}(A,G)$. Then χ extends to G as an irreducible representation of G satisfying relation (5.12.i).

Proof. By theorem (1.5), there is a cyclic representation U of G extending χ. Since G has a countable base for its topology, U is a direct integral of representations $U^{(\alpha)}$ with respect to some locally compact space X and some positive nonzero Borel measure ι on X, in such a way that ι-almost all $U^{(\alpha)}$ are irreducible and nonzero (see Godement [2]). Let f be a vector field in $L^2_{\Lambda,\iota}$. Then we have for $x \in A$ and ι-almost all $\alpha \in X$

$$(U_x f)(\alpha) = U_x^{(\alpha)} f(\alpha).$$

Theorem (5.12) shows that for $x \in A$ and ι-almost all α we also have

$$(U_x f)(\alpha) = \chi(x) f(\alpha).$$

Hence we see for all $x \in A$ that

$$\iota\left(\{\alpha \in X : U_x^{(\alpha)} f(\alpha) \neq \chi(x) f(\alpha)\}\right) = 0.$$

Now, by Godement [2], there is a sequence $\{f_n : n \geq 1\}$ in $L^2_{\Lambda,\iota}$ such that for each $\alpha \in X$ the set $\{f_n(\alpha) : n \geq 1\}$ is dense in $H^{(\alpha)}$. There is also a sequence $\{x_m : m \geq 1\}$ that forms a dense subset of A. Because of the countable additivity of ι, we have

$$\iota\left(\left\{\alpha \in X : \text{ for some } m,n \geq 1,\ U_{x_m}^{(\alpha)} f_n(\alpha) \neq \chi(x_m) f_n(\alpha)\right\}\right) = 0$$

and so for ι-almost all α and all $m, n \geq 1$ we have

$$U_{x_m}^{(\alpha)} f_n(\alpha) = \chi(x_m) f_n(\alpha).$$

Since for each α the set $\{f_n(\alpha) : n \geq 1\}$ is dense in $H^{(\alpha)}$, and since the set $\{x_m : m \geq 1\}$ is dense in A, for ι-almost all α we actually have

$$U_x^{(\alpha)}\xi = \chi(x)\xi \quad \text{for} \quad x \in A \quad \text{and} \quad \xi \in H^{(\alpha)}.$$

Thus for ι-almost all α, $U^{(\alpha)}$ extends χ. Because ι-almost all $U^{(\alpha)}$ are irreducible, and since $\iota(X) > 0$, the proof is complete.

Remark. Theorems (5.12) and (5.13) show that our use of the word "extension" coincides with that of Grosser and Moskowitz [2,section 5, Definition 2], provided certain conditions are present. Without these conditions, our extensions are not necessarily "extensions" in the sense of Grosser and Moskowitz [2]. For example, let $A = G = \mathbb{R}$, and let χ and χ_1 be two unequal characters of A, regarded as 1-dimensional representations of A. The 2-dimensional representation $U = \chi \oplus \chi_1$ of G clearly extends χ, but χ and U do not satisfy (5.12.i), and hence U is not an "extension" of χ in the sense of Grosser and Moskowitz [2].

(5.14) Theorem. Let G be a locally compact group, and let A be a closed subgroup of the centre of G. Then A is an extending subgroup of G if and only if every character of A belongs to $\underline{\underline{P}}(A,G)$.

Proof. To prove the nontrivial portion of this statement, we shall consider an arbitrary cyclic representation U of A and show that if the stated condition holds then U has an extension defined on G.

Let X denote the character group of A. According to Stone's theorem (Hewitt and Ross [2,(33.8)]), there is a measure $\iota \in M^+(X)$ such that U is unitarily equivalent to the representation $y \longmapsto V_y$, V_y being the operator on $L^2(X,\iota)$ defined by setting

$$(V_y h)(\chi) = \chi(y)h(\chi)$$

for all $\chi \in X$ and $h \in L^2(X,\iota)$. Each character χ we shall regard as a 1-dimensional irreducible representation of A, and we shall write $H(\chi)$ for the representation space of χ.

Now, for all $\chi \in X$, χ by assumption belongs to $\underline{\underline{P}}(A,G)$, and since A is contained in the centre of G, χ is invariant. Thus by (5.12) there is a cyclic representation $U^{(\chi)}$ of G that extends χ in such a way that we have

$$U_x^{(\chi)} = \chi(x)I_\chi \quad \text{for} \quad x \in A,$$

where I_χ denotes the identity operator on $H(U^{(\chi)})$. Since the conditions of (5.11) are satisfied by G, χ, ι, and the family $\{U^{(\chi)} : \chi \in X\}$, we may form the direct integral $U' = \int_X \oplus U^{(\chi)} d\iota$, with representation space $L^2_{\Lambda,\iota}$.

The space $L^2(X,\iota)$ may be regarded as a subspace of $L^2_{\Lambda,\iota}$ as follows. Each $h \in L^2(X,\iota)$ we may regard as an element of $L^2_{\Lambda,\iota}$ since each of its values $h(\chi)$ is an element of the subspace $H(\chi)$ of $H(U^{(\chi)})$, and the norm of h so regarded is clearly identical with its norm as an element of $L^2(X,\iota)$.

For each $h \in L^2(X,\iota)$ and each $x \in A$ we have

$$\begin{aligned}(U_x'h)(\chi) &= U_x^{(\chi)}h(\chi)\\ &= \chi(x)h(\chi)\\ &= (V_xh)(\chi)\end{aligned}$$

for ι-almost all $\chi \in X$. Thus U' is an extension of V to G. Since U and V are equivalent, U also has an extension to G, as was desired.

We now derive two different sets of conditions under which a character χ of A extends to G. These conditions appear as (5.15) and (5.18) <u>infra</u>. Each may be regarded as a partial converse of Douady's theorem. The first of these is well known in the case of a discrete group G, when it is sometimes called Mackey's theorem (see Curtis and Reiner [1,(51.15)]).

(5.15) Theorem. Let A be a closed abelian normal subgroup of a locally compact group G, and let χ be a character of A. Suppose that the two following conditions hold.

(i) The subgroup $F(\chi)$ of G is open.

(ii) There is a closed subgroup H of $F(\chi)$ such that $AH = F(\chi)$, $A \cap H = \{e\}$, and $F(\chi)$ is the topological semi-direct product of A and H. Then χ extends to G as a continuous positive-definite function.

Proof. It will suffice to show that we have $\chi \in \underline{\underline{\underline{P}}}(A,F(\chi))$. We shall show in fact that χ extends to $F(\chi)$ as a character of $F(\chi)$. Indeed, each element x of G may be written uniquely in the form $x = ah$ with $a \in A$ and $h \in H$. Thus we may define a function χ' on AH by setting

$$\chi'(ah) = \chi(a).$$

The function χ' takes values in the circle group. It is a homomorphism since for a and a' in A and h and h' in H we have

$$\begin{aligned}\chi'((ah)(a'h')) &= \chi'(a(ha'h^{-1})hh') \\ &= \chi(aha'h^{-1}) \\ &= \chi(a)\chi(ha'h^{-1}) \\ &= \chi(a)\chi(a') \\ &= \chi'(ah)\chi'(a'h'),\end{aligned}$$

where we have used the fact that $h \in H \subseteq F(\chi)$. Finally χ' is continuous, since the topology for $F(\chi)$ is the topology obtained by regarding $F(\chi)$ as the product $A \times H$. Thus if $\{x_\alpha : \alpha \in D\}$ is a net in $F(\chi)$ such that

$\lim_\alpha x_\alpha = e$, and if for each $\alpha \in D$ we write $x_\alpha = a_\alpha h_\alpha$ with $a_\alpha \in A$ and $h_\alpha \in H$, then we have $\lim_\alpha a_\alpha = e$. Hence we have

$$\lim_\alpha \chi'(x_\alpha) = \lim_\alpha (a_\alpha) = 1.$$

This completes the proof.

(5.16) Definition. Let G be a topological group. If the centre $Z(G)$ of G is an extending subgroup, we shall say that G has the central extension property.

Clearly, if G has the central extension property, then every closed subgroup of $Z(G)$ is extending, because $Z(G)$ is abelian and has the extension property.

It is also clear from (5.6) *supra* that all abelian closed normal subgroups of a connected locally compact group G are extending if and only if all such subgroups are contained in $Z(G)$ and $Z(G)$ is extending. Thus for connected locally compact groups, at least, the central extension property is quite important.

A fairly large class of groups have the central extension property, as we shall now show. First we prove a preliminary result.

(5.17) Theorem. Let G be a locally compact group with an open subgroup H, and suppose that H has the central extension property. Then G has the central extension property.

Proof. It will suffice to show that every character on $Z(G)$ extends to G as a positive-definite continuous function, in view of Theorem (5.14).

Now, the subgroup $Z(G) \cap H$ is contained in $Z(H)$, and therefore the restriction χ_o of χ to $Z(G) \cap H$ extends to H as a representation

$U^{(o)}$ of H, in such a way that for $z \in Z(G) \cap H$ we have

$$U_z^{(o)} = \chi_o(z)I$$

where I is the identity operator on $H(U^{(o)})$ (see (5.12) supra). Let $G_1 = Z(G)H$. Then G_1 is open since it contains H, and is a subgroup since $Z(G)$ is normal in G.

If z and z' belong to $Z(G)$ and h and h' belong to H, and if we have $zh = z'h'$, then we have $z^{-1}z' = h(h')^{-1}$, and so both $z^{-1}z'$ and $h(h')^{-1}$ are in $Z(G) \cap H$. Thus we have

$$U_h^{(o)}U_{(h')^{-1}}^{(o)} = U_{h(h')}^{(o)} = \chi(h(h')^{-1})I = \chi(z^{-1}z')I = \chi(z)^{-1}\chi(z')I$$

and so we see that

$$\chi(z)U_h^{(o)} = \chi(z')U_{h'}^{(o)} .$$

The value of $\chi(z)U_h^{(o)}$, then, depends only on the element zh of G_1 and not upon the way it is factored with respect to $Z(G)$ and H. Hence we may define (for $zh \in G_1$ with $z \in Z(G)$ and $h \in H$)

$$U_{zh} = \chi(z)U_h^{(o)}.$$

If z' and h' are also elements of $Z(G)$ and H, respectively, then we have

$$\begin{aligned} U_{(zh)(z'h')} &= U_{zz'hh'} \\ &= \chi(zz')U_{hh'}^{(o)} \\ &= \chi(z)\chi(z')U_h^{(o)}U_{h'}^{(o)} \\ &= \chi(z)U_h^{(o)}\chi(z')U_{h'}^{(o)} \\ &= U_{zh}U_{z'h'} . \end{aligned}$$

Thus we see that $y \mapsto U_y$ is a homomorphism of G_1 into the unitary group of $H(U^{(o)})$. For $z \in Z(G)$, we clearly have $U_z = \chi(z)I$. Thus we will have finished if we show that U is a continuous representation of G_1.

In each left coset of H in G_1, choose an element from the group $Z(G)$, thus forming a set $Z_o \subseteq Z(G)$ such that every left coset of H in G has the form zH for a <u>unique</u> $z \in Z_o$. Let $z \in Z_o$, let $zh = x \in zH$, and let $\{x_\alpha : \alpha \in D\}$ be a net in zH whose limit is x. Each x_α may be written in the form zh_α, for some $h_\alpha \in H$, and we have

$$\begin{aligned} \lim_\alpha h_\alpha &= \lim_\alpha (z^{-1}x_\alpha) \\ &= z^{-1}x \\ &= h. \end{aligned}$$

Therefore, we have for ξ and η in $H(U)$,

$$\begin{aligned} \lim_\alpha < U_{x_\alpha}\xi, \eta > &= \lim_\alpha < \chi(z)U^{(o)}_{h_\alpha}\xi, \eta > \\ &= \lim_\alpha < U^{(o)}_{h_\alpha}(\chi(z)\xi), \eta > \\ &= < U^{(o)}_h(\chi(s)\xi), \eta > \\ &= < \chi(z)U^{(o)}_h\xi, \eta > \\ &= < U_x\xi, \eta > . \end{aligned}$$

This shows that U is continuous when restricted to the coset zH of H. Since G_1 is a disjoint union of the zH, each of which is open, this entails continuity of U. Thus the character χ of $Z(G)$ extends to $Z(G)H$ as a positive-definite continuous function, by Theorem (1.7). Since $Z(G)H$ is an open subgroup of G, χ can thus be extended to all of G.

(5.18) <u>Theorem</u>. Let G be a locally compact group whose left and right uniform structures are equivalent. Then G has the central extension property.

<u>Proof</u>. The hypothesis is equivalent to the "small invariant neighborhood property" of Grosser and Moskowitz [3]. For the left and right uniform structures of G are equivalent if and only if for every open neighborhood U of e, there is an open neighborhood V of e such that $xVx^{-1} \subseteq U$ for all $x \in G$ (see Hewitt and Ross [1,(4.14.g)]). For this it is obviously sufficient that U contain an invariant neighborhood of e. Conversely, if this condition holds then the set $V^* = \bigcup_{x \in G} (xVx^{-1})$ is clearly an invariant open neighborhood of e contained in U.

Therefore, we may use the structure Theorem (2.13) of Grosser and Moskowitz [3] which implies that a locally compact group G with the small invariant neighborhood property has an open subgroup H of the form $V \times K$, where V is a locally compact abelian group and K is a compact group. By Theorem (5.17), we need only show that H has the central extension property.

Let χ be a character of $Z(H)$, and let χ_o denote the restriction of χ to $Z(H) \cap K$. Since K is compact, K has the extension property, and since $Z(H) \cap K \subseteq Z(K)$, χ_o is an invariant character of $Z(H) \cap K$ in K. Therefore, by (5.12), χ_o has a cyclic extension $U^{(o)}$ to K for which we have

$$U_x^{(o)} = \chi_o(x)I \quad \text{for} \quad x \in Z(H) \cap K,$$

where I denotes the identity operator on $H(U^{(o)})$.

Let χ_1 denote the restriction of χ to V (clearly $V \subseteq Z(H)$). Let $U : vk \longmapsto \chi_1(v)U_k^{(o)} = U_{vk}$ be the tensor product of χ_1 and $U^{(o)}$. Then U is a representation of $H = V \times K$. Let $z \in Z(H)$, and let $z = vk$, with

$v \in V$ and $k \in K$. Then we have $k = v^{-1}z$, so that $k \in Z(H)$, since $V \subseteq Z(H)$. It follows that $k \in Z(H) \cap K$, and so we have

$$\begin{aligned} U_z &= \chi_1(v)U_k^{(o)} \\ &= \chi_1(v)\chi_o(k)I \\ &= \chi(v)\chi(k)I \\ &= \chi(z)I, \end{aligned}$$

which shows that U extends χ to H. By Theorem (1.7), we, therefore, have $\chi \in \underline{\underline{P}}(Z(H),H)$, and so by Theorem (5.14) $Z(H)$ is extending.

(5.19) <u>Corollary</u>. If V is a locally compact abelian group and K is a compact group, then $V \times K$ has the central extension property.

In view of this corollary, and the fact that V and K both have the extension property, it seems reasonable to make the following conjecture.

(5.20) <u>Conjecture</u>. If V is a locally compact abelian group and K is a compact group, then $V \times K$ has the extension property.

(5.21) <u>Example and Discussion</u>. Let G be as in (5.10). Then G has the central extension property.

Firstly, the centre of G is the set of matrices $\begin{bmatrix} 1 & 0 & 0 \\ 0 & 1 & 0 \\ w & 0 & 1 \end{bmatrix}$ with $w \in \mathbb{R}$, as is readily checked. Thus every character χ of $Z(G)$ has the form

$$\chi : \begin{bmatrix} 1 & 0 & 0 \\ 0 & 1 & 0 \\ w & 0 & 1 \end{bmatrix} \longmapsto \exp(i\,\alpha\, w)$$

for some $\alpha \in \mathbb{R}$. For each $\alpha \in \mathbb{R}$, Dixmier [1,Formula (2)] gives us a representation $U^{(\alpha)}$ of G, acting on $L^2(\mathbb{R})$, whose definition follows.

For $f \in L^2(\mathbb{R})$, $x = \begin{bmatrix} 1 & 0 & 0 \\ u & 1 & 0 \\ w & v & 1 \end{bmatrix} \in G$, and $v' \in \mathbb{R}$ we set

$$U_x^{(\alpha)} f(v') = f(v'+v)\exp[i\alpha(v'u+w)].$$

If $x = \begin{bmatrix} 1 & 0 & 0 \\ 0 & 1 & 0 \\ w & 0 & 1 \end{bmatrix} \in Z(G)$, then we have

$$U_x^{(\alpha)} f(v') = f(v')\chi(x),$$

showing that $U^{(\alpha)}$ extends χ. Therefore, by (5.14), G has the central extension property. This example is of interest for the reason that G does not have the extension property (see (5.10) supra).

We conclude this entire investigation with a conjecture, prompted by (5.18) and (5.21).

(5.22) Conjecture. Every locally compact group has the central extension property.

(5.23) Equivalent Conjecture (Converse of (5.3)). For every locally compact group G with closed abelian normal subgroup A, and for each character χ on A, if $F(\chi)$ is open in G, then $\chi \in \underline{\underline{P}}(A,G)$.

Proof of the equivalence. If G is a locally compact group, and χ is a character of $Z(G)$ that does not belong to $\underline{\underline{P}}(Z(G),G)$, then (5.23) is clearly false, since $F(\chi) = G$ is open in G.

On the other hand, suppose that (5.22) is true, and let G, A and χ be as in (5.23), with $F(\chi)$ open in G. Then the kernel $\ker(\chi)$ of χ is a closed normal subgroup of $F(\chi)$ contained in A, and the group $F_1 = F(\chi)/\ker(\chi)$ contains the subgroup $A/\ker(\chi)$ in its centre. The character χ defines a character χ_1 on $A/\ker(\chi)$ in an obvious way, and according to the hypothesis that (5.22) is true, we have an extension $\varphi_1 \in \underline{\underline{P}}(F(\chi)/\ker(\chi))$

for χ_1. The function φ_1 defines in the natural way a positive-definite function φ on $F(\chi)$, constant on cosets of $\ker(\chi)$, and extending χ. Since $F(\chi)$ is open in G, we therefore have $\chi \in \underline{\underline{P}}(A,G)$ as desired.

BIBLIOGRAPHY

Akutowitz, E. J.: [1] On extrapolating a positive-definite function from a finite interval. Math. Scand. 7 (1959), pp. 157-169.

Alfsen, E., and P. Holm: [1] A note on compact representations and almost periodicity in topological groups. Math Scand. 10 (1962), pp. 127-136.

Bourbaki, N.: [1] Elements of Mathematics: General Topology. Part 1. Hermann-Addison-Wesley, 1966.

__________: [2] Elements of Mathematics: General Topology. Part 2. Hermann-Addison-Wesley, 1966.

Chover, J.: [1] On normalized entropy and the extension of a positive-definite function. J. Math. Mech. 10 (1961), pp. 927-945.

Curtis, C. W., and I. Reiner: [1] Representation Theory of Finite Groups and Associative Algebras. Interscience Publishers, Inc., New York, 1962.

Devinatz, A.: [1] On the extension of positive-definite functions. Acta Math. 102 (1959), pp. 109-134.

Dixmier, J.: [1] Sur les représentations unitaires des groupes de Lie nilpotents. IV. Can. J. Math. 11 (1959), pp. 321-344.

Eymard, P.: [1] L'algèbre de Fourier d'un groupe localement compact. Bull. Soc. Math. France 92 (1964), pp. 181-236.

__________: [2] Algèbres A_p et convoluteurs de L^p. Seminaire Bourbaki, 22 ème année (1969-1970), No. 367.

Godement, R.: [1] Les fonctions de type positif et la théorie des groupes. Trans. Amer. Math. Soc. 63 (1948), pp. 1-84.

__________: [2] Sur la théorie des représentations unitaires. Ann. of Math. (2) 53 (1951), pp. 68-124.

Grosser, S., and M. Moskowitz: [1] On central topological groups. Trans. Amer. Math. Soc. 127 (1967), pp. 317-340.

__________: [2] Representation theory of central topological groups. Trans. Amer. Math. Soc. 129 (1967), pp. 361-390.

__________: [3] Compactness conditions in topological groups. I. Columbia University Department of Mathematics preprint.

__________: [4] Compactness conditions in topological groups. II. Columbia University Department of Mathematics preprint.

Hewitt, E.: [1] Linear functionals on almost periodic functions. Trans. Amer. Math. Soc. 74 (1953), pp. 303-322.

Hewitt, E., and K. A. Ross: [1] Abstract Harmonic Analysis, vol. I. Springer, Berlin, 1963.

________: [2] Abstract Harmonic Analysis, vol. II. Springer, Berlin, 1970.

Kakutani, S., and K. Kodaira: [1] Über das Haarsche Mass in der lokal bikompakten Gruppe. Proc. Imp. Acad. Tokyo 20 (1944), pp. 444-450.

Kaniuth, E.: [1] Zur harmonischen Analyse klassenkompakter Gruppen. Math. Zeit. 110 (1969), pp. 297-305.

Kreĭn, M. G.: [1] Sur le problème du prolongement des fonctions hermitiennes positives et continues. Dokl. Akad. Nauk SSSR, N.S. 26 (1940), pp. 17-22.

________: [2] A ring of functions on a tpological group. Dokl. Akad. Nauk SSSR, N.S. 29 (1940), pp. 275-280.

________: [3] On a special ring of functions. Dokl. Akad. Nauk SSSR, N.S. 29 (1940), pp. 355-359.

Mackey, G. W.: [1] On induced representations of groups. Amer. J. Math. 73 (1951), pp. 576-592.

________: [2] Induced representations of locally compact groups. I. Ann. of Math. (2) 55 (1952), pp. 101-139.

Murakami, S.: [1] Remarks on the structure of maximally almost periodic groups. Osaka Math. J. 2 (1950), pp. 119-129.

Reiter, H.: [1] Sur la propriété (P_1) et les fonctions de type positif. C. R. Acad. Sci. Paris 258 (1964), pp. 5134-5135.

________: [2] On some properties of locally compact groups. Nederl. Akad. Wetensch. Proc. Ser. A 68 = Indag. Math. 27 (1965), pp. 697-701.

Rudin, W.: [1] An extension theorem for positive-definite functions. Duke Math. J. 37 (1970), pp. 49-53.

________: [2] The extension problem for positive-definite functions. Ill. J. Math. 7 (1963), pp. 532-539.

Schenkman, E.: [1] Group Theory, Van Nostrand, Princeton, 1965.